Wireless Sensor Networks: Architectures, Protocols and Applications

Wireless Sensor Networks: Architectures, Protocols and Applications

Edited by **Sharon Garner**

WILLFORD PRESS

New York

Published by Willford Press,
118-35 Queens Blvd., Suite 400,
Forest Hills, NY 11375, USA
www.willfordpress.com

Wireless Sensor Networks: Architectures, Protocols and Applications
Edited by Sharon Garner

International Standard Book Number: 978-1-68285-122-7 (Hardback)

Printed in the United States of America.

Contents

Preface

This book is a compilation of topics related to wireless sensor networks, wireless architecture and protocols, their applications, etc. It provides a detailed overview of the present status of WSNs. Some of the chapters included herein provide significant information on topics like power efficiency and energy consumption in wireless sensor networks, designing architecture for wireless sensor networks, routing protocols for WSNs, etc. It is an essential guide for students and academicians engaged in this field.

The information shared in this book is based on empirical researches made by veterans in this field of study. The elaborative information provided in this book will help the readers further their scope of knowledge leading to advancements in this field.

Finally, I would like to thank my fellow researchers who gave constructive feedback and my family members who supported me at every step of my research.

Editor

LOAD BALANCING IN WIRELESS AD-HOC NETWORKS WITH LOW FORWARDING INDEX

Reena Dadhich[1] & Aditya Shastri[2]

[1]Department of MCA, Engineering College Ajmer, India
reena.dadhich@gmail.com
[2]Department of Computer Science, Banasthali University, India.
adityashastri@yahoo.com

Abstract

A wireless ad-hoc network comprises of a set of wireless nodes and requires no fixed infrastructure. For efficient communication between nodes, ad-hoc networks are typically grouped in to clusters, where each cluster has a clusterhead (or Master). In our study, we will take a communication model that is derived from that of BlueTooth. Clusterhead nodes are responsible for the formation of clusters each consisting of a number of nodes (analog to cells in a cellular network) and maintenance of the topology of the network. Consequently, the clusterhead tend to become potential points of failures and naturally, there will be load imbalanced. Thus, it is important to consider load balancing in any clustering algorithm. In this paper, we consider the situation when each node has some load, given by the parameter forwarding Index.

Keywords

Ad-hoc Networks, Clustering, Bluetooth, Forwarding Index,Algorithm.

1. INTRODUCTION

Ad-hoc networks are expected to play a significant role in the future mobile computing applications. A wireless ad-hoc network consists of a set of self-organizing mobile nodes which required no fixed infrastructure and which communicate with each other over wireless links. For efficient communication between nodes, ad-hoc networks are typically grouped into clusters, where each cluster has a clusterhead (or Master). Communication between nodes in different clusters is through gateway nodes; these are also known as bridge nodes. Bluetooth is an emerging technology for indoor wireless picocellular environment and it employs a master-slave model for communication between nodes. In this model each cluster has a star topology, with a master at the center of the star, and the Master controls the traffic to the Slaves. In order to streamline flow of information between nodes and to adapt to topological changes, the entire network is divided into cluster of nodes.

Efficient clustering and topology construction algorithms play a very important role in the fast connection establishment of ad-hoc networks. The performance of these networks is chiefly dependent on the device discovery time, i.e. the time taken by a node to discover and to connect to another node in its radio range which is already part of the existing network. This device discovery time is also crucial in other situations. For example, when a large number of devices within radio range of each other are powered on, the time taken to complete the formation of the network is an important performance criterion.

Throughout the paper we work with Bluetooth model which is a synchronous system in which every node has a unique Id, but does not know the ID of any other node. In this paper we study the Randomized algorithm for cluster formation, due to Aggarwal et. al. [9] for asynchronous complete networks of N nodes (all within radio range of each other), which can be used to construct a minimal set of star-shaped clusters of limited size having the forwarding index within at most a constant factor of the optimal.

Most of the existing algorithms focus on partitioning the network into clusters and differ mainly in the clusterhead election criterion. Further, they also do not consider the stability of the network while clustering. As a result, these nodes take greater responsibility and thus their energy gets depleted faster making them to drop out.   This situation creates congestion in the network because large number of routes passes through the clusterhead node.

Thus, the clustering algorithm should guarantee that the extra workload is always balanced between all the nodes of the network. It other words, the responsibility of acting as clusterheads should be fairly distributed in the network. It will also be bad from the fault-tolerance point of view for if such a node were to fail a large part of the network would come to a halt. Therefore, there is a need to evenly distribute the routing [9] & load among all nodes in the cluster (i.e., load balancing) [1].

In this paper, we propose a parameter; called forwarding Index for ad-hoc networks. Forwarding Index is a measure of routing to be evenly distributed proposed in [2]. The vertex (edge) forwarding index of the network is the minimum value of the largest load occurring at a vertex (edge) taken over all routings, where load of a vertex (edge) is defined as the number of routes passing through that vertex or edge. The problem of determining the forwarding Index of a network is NP-complete [2,3,6].

In ad-hoc networks the value of forwarding Index for clusterhead node is maximum. For load balancing in wireless ad-hoc networks we propose a design for distributing the load of clusterhead node among under loaded nodes by means of calculating the forwarding Index parameter. By distributing the load of clusterhead we will be able to minimize the average execution time and maximize the lifetime of the overloaded (clusterhead) node.

1.1. RELATED WORK

For the wireless ad-hoc networks, the problem of load balancing is also defined by many authors. We have looked many papers [11, 12, 13, 1, 2] that deal with the problem of load balancing in ad-hoc networks. Reach of these papers attacks the problem of load balancing in different ways. The paper by S.K DAS et. al. [14] tries to devote the load balancing algorithm that would try to find the best node to share the load with while minimizing the communication overhead involved in load – balancing. The another paper by D. Turgut et. al. [2] deals with ad-hoc networks where mobile nodes have been loosely classified into clusters based on their current location. The authors proposed a load – balancing heuristic to extend the battery life of a clusterhead before allowing it to retire and allow another node to become the clusterhead.

2. CLUSTERING MODEL OF WIRELESS AD-HOC NETWORKS

Wireless communication network can be presented by an undirected graph G=(V, E), where V and E are the set of vertices and edges respectively for graph G. Each node $v \in V$ represents a wireless station and every undirected edge $e \in E$ defines a neighbor relationship between two nodes, that, is to say it indicates two nodes at the end of edge those can communicate with each other, for any nodes u and v ∈ V, if v is an adjacent node of neighbors of u and is not a neighbor of u, we say v is two hop node of u.

Each node has a unique identifier (ID) number and only has a transceiver operated in half-duplex mode. Node requires adjacent nodes and all the topology connection of the whole network through transmitting or receiving default control packet or message. At first every node sends its control packet to indicate its existence. When a node receives a control packet of adjacent node, it updates its related data table. And when it transmits its control packet once again, the packet includes all the information of its adjacent nodes it knew. From this we can see that if all nodes in network transmit a control packet respectively, every node will know all its neighbors. If all nodes transmit another control packet again, the node will know its two hop nodes. With further exchange, every node will know the topology of the wireless network. Based on the locality information, distributed control is implemented.

3. CLUSTERING ALGORITHMS

One of the fundamental problems in wireless mobile computation is efficient cluster formation. It is well known that optimal cluster formation is NP-complete.

In this section we study the problem of distributed cluster formation in a wireless ad-hoc environment. We work on Bluetooth model which is a synchronous system in which every node has a unique ID, but does not know the ID of any other node. A node trying to discover other nodes broadcasts a generic message and does not advertise its ID in the massage. The replying node gives its ID in the reply message, but does not know which node it is replying to. However, after a device has discovered another device, much more information can be passed between them with relatively less overheads. This model is ideally suited to frequency hopping systems, where the devices hop on a sequence of frequencies, and the messages are repeatedly broadcast in order to reach other nodes.

The bluetooth SIG aims to provide solutions for short-range wireless connectivity between pervasive devices, like PDAs, mobile phones, palmtops, laptops, pagers, etc. It is meant to be cable-replacement solution for desktops, keyboards and other peripherals devices. The potential applications range from smart home appliances to wireless connectivity to backbone data networks. Bluetooth is being considered for use by the top players in the consumer electronics market. Products would include wireless headsets, cameras, watches and portable games.

The automotive industry is also looking to use Bluetooth technology as the key solution for onboard wireless communication systems, connecting vehicular and external networks. These and other applications in the office and classroom environments, like shared whiteboards, would make it important for the devices to quickly self-organize into an ad-hoc network.

4. A RANDOMIZED ALGORITHM FOR CLUSTER FORMATION

In this section, closely following Aggarwal et.al.[9], we describe a two-stage algorithm for partitioning the set of nodes into a connected set of star-shaped clusters, while keeping the size of the clusters at their maximum. An important idea used in this algorithm is to make a device continuously broadcast or continuously listen, in order to increase the probability of the message reaching another device.

The first stage of algorithm is randomized, at the end of which each node either becomes a Master-designate or a Slave-designate. For a network of N nodes and maximum cluster size S, the ideal number of Masters is k=[N/(S+1)]. The second stage uses a deterministic algorithm to decide on the final set of Masters and Slaves, and to efficiently assign Slaves to Masters. A Super-master is elected, which is required for counting the actual number of Masters, and for collecting information about all the nodes. This stage also corrects the effects of the randomness introduced in the previous stage. The election of the super-Master is interleaved with the cluster formation, which speeds up the ad-hoc network formation. The super-master can then run any centralized algorithm to form a network of desired topology.

In the following discussion, we use the following terminology:

➢ Slave-designate: a node which did not succeed in any of the Bernoulli trials, and is not yet part of any cluster.
➢ Slave: a Slave-designate which becomes part of a cluster.
➢ Master-designate: a node which had a successful Bernoulli trial, and has not yet collected Inquiry response from enough slave-designates and has not timed out (CLUSTER_TO).
➢ Master: a node which has collected response from S Slave-designates or has timed out (CLUSTER_TO).
➢ Proxy-slave: a Slave which has been identified by its Master to participate in the super-master election on its behalf.
➢ Super-master-designate: a Master which has collected k responses from other cluster, or has reached the SUPERM_TO.
➢ Super-master: a Master which has response from all other clusters, and has information about all the nodes in the network.

The algorithm given by Aggarwal et.al. [9] is described below:

Stage 1: Each of the N nodes conduct T rounds of a Bernoulli trials with probability of success equal to p. A node which is successful at least once becomes a Master-designate and the remaining nodes become slave-designate.

Stage 2: We make the following additional assumptions on the various timeout value used by the nodes. These timeouts are the same for all the nodes.

Assumption-1: Each node has a CLUSTER_TO value such that if it inquires for this period of time, and there are enough number of nodes in its radio range which are scanning for Inquiry packets, then at least S devices will respond to it.

Assumption - 2: Each node has a SUPERM_TO value such that a node inquiring for this period receives responses from at least 2k nodes that are scanning. For practical purposes, we assume that $P[X > 2k]$ is very small, for reasonably large k.

Assumption-3: A set inquiring nodes catch one scanning node each, well before the SUPERM_TO period.

Algorithm 4.1[1]: Master-designates and slave-designates are using state-1, as described above. Let X be the actual number of master-designated.

Each master-designate inquires continuously until neither the CLUSTER_TO nor the SUPERM_TO is reached. If a response is received from a Slave designate, it made a Slave in its cluster by paging and making a connection to it, as long as the maximum cluster size is not exceeded. If this is the first Slave of the cluster, the Master-designate instructs it to become a proxy-slave.

When the cluster becomes full, the Master-designate declares itself master, and any future inquiry responses from Slave-designates are ignored. As part of the Super-master election which is interleaved with the cluster formation, the Master/Master-designate collects up to k responses from Proxy-slaves of other clusters, or times out (SUPERM_TO), whichever is earlier. At this point, the node declares itself Super-master-designate.

However , this happens only after CLUSTER_TO has occurred. If the Master-designate has not collected any responses by the CLUSTER-TO period, then it becomes a Slave-designate and starts scanning. Node have unique Ids, and the master with the highest Id is chosen to the Super-master, there is exactly one Super master that is elected. A slave designate continuously scans for Inquiry massages. If, on sending an inquiry response, the inquirer does not page it and establish a connection, then it goes back to scan state. However, If the inquirer connects to it, then it becomes a Slave of the Cluster headed by its inquirer, and stops scanning. If the Master/Master-designate directs it to become a Proxy-slave, it goes into scan for the Super-master election.

5. FORWARDING INDEX AND WIRELESS AD – HOC NETWORKS

Our main objective is to study the problem of evenly distributed cluster formation in ad-hoc wireless environment. It is desirable to have these clusters as evenly distributed as possible over the network to avoid congestion in the network. Clusterhead form a virtual backbone and may be used to route packets (messages) for nodes in their cluster. Nodes are assumed to have non-deterministic mobility pattern. Diffusing node identities along the wireless links forms clusters. Different heuristics employ different policies to elect clusterheads. Several of these policies are biased in favor of some nodes. As a result, these nodes shoulder greater responsibility and may deplete their energy faster causing them to drop out of the network (i. e. there occurs a congestion in the network). Therefore, there is a need to minimize the load of clusterhead. Clusterheads maintain cluster databases for routing purposes.

To avoid the congestion in the network we propose the concept of forwarding index for the clusterhead of the cluster. The clusterhead-forwarding index of the network (cluster) is the minimum value of the largest load occurring at a clusterhead taken over all nodes in the cluster, where load of a clusterhead is defined as the number of paths (routes) passing through that clusterhead. This helps to evenly distribute the responsibility of acting as clusterheads among all nodes to avoid congestion in the network. This congestion is also bad from the fault-tolerance point of view for if clusterhead of such a cluster were to fail a large part of the network come to a halt. Computing forwarding index of general network was shown to be NP-complete [6] by Saad [3] and problem of optimal clustering is also NP-complete.

In a communication network data, messages, etc., are transmitted from each node to any other node. A convenient way to achieve this is to have for every source node a designated route, a sequence of intermediate nodes for every destination. A set of nodes (which are processors or communication centers), with links between some of them for the purposes of communicating data or messages is usually represented by graphs. Generally the nodes are to be interpreted as computer/communication devices. In practice, the networks to be constructed may range from arrays of microcomputers to systems of large geographically remote centers. Instead of speaking of nodes and links we speak of vertices and edges.

The network connecting the n nodes is designed by specifying first the bi-directional communication lines or channels, i.e., those pairs of nodes having direct communication. Interconnection is limited by a port constraint d ≥ 2 common to each node; i.e., at most d (where d is the degree of the graph) communication lines can be attached to any node. Since it follows in general that not every pair of nodes will have direct communication, the network design must also specify a set of n (n-1) paths called a *routing,* indicating for each *x* and *y* ≠ *x* the path or fixed sequence of lines which carries the data transmitted from node *x* to node *y*. Implicit here is that in addition to being data sources and sinks, the nodes can serve a forwarding function for the data being communicated between other nodes. Note that, generally, the path from node *x* to node *y* need not be the reverse of the path from node *y* to node *x*.

If some nodes or links fail, it is important to know which paths of the network are destroyed, and quite naturally it seems a 'good' routing should not load any vertex or edge too much, in the sense that not too many paths of the routing should go through it. In order to measure the load of a vertex, Chung, Coffman, Reiman and Simon introduced in [2] the notion of Forwarding Index [2,3,10].

The network *forwarding index* ξ, defined as the maximum number of paths passing through any node, i.e., ξ is the maximum forwarding being done in the network. With n and d given we consider the specific problem of finding networks that minimize the forwarding index; we call this *forwarding index problem.*

Fig. 1 shows an example for n = 6 and d = 3. According to the routing indicated, nodes 1 and 4 forward the traffic on one path each; nodes 5 and 6 forward the traffic on two paths each; and nodes 2 and 3 forward the traffic on a total of four paths each. Thus ξ = 4 for this network.

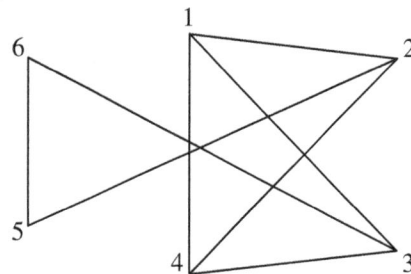

n=6,d=5

Routing:
1. Both Paths between
    1 & 5 through 2
    4 & 5 through 2
    1 & 6 through 3
    4 & 6 through 3
    2 & 6 through 5
    3 & 5 through 6
2. Paths from 2 to 3 pass through 4
3. All other communications are direct

Figure 1.

Concrete applications of the forwarding index problem can be found in problems of maximizing network capacity. For example, assume symmetric transmission requirements in the sense that the transmission rate, say λ, is the same for each node to every other node. The total rate at which data originates and terminates at each node is, therefore 2(*n*-1) λ, and the total transmission rate among the nodes is *n* (*n*-1) λ.The amount of forwarding at a node is assumed to be limited by a capacity *c* common to all nodes. Specifically, the local transmission rate at a node 2(*n*-1) λ, plus the rate at which it forwards data for other nodes cannot exceed *c*. In Fig. 1, for example, since the nodes 2 and 3 forwards the most traffic and since the traffic at these nodes is 2(*n*-1) λ + 4λ = 14λ, we must have *c* ≥ 14λ. The constraint on node capacity requires that $(2n-1) \lambda + \xi \lambda \leq c$. The local traffic originating or terminating at each node must, therefore, satisfy

$$(2n-1) \lambda \leq 2(n-1)c/\xi + 2(n-1) \qquad \ldots\ldots\ldots(1)$$

thus defining an *effective* node capacity $c/\xi + 2(n-1)$. The corresponding bound on the total data transmission rate defines the network capacity

$$n(n-1) \lambda \leq (nc/2)/(1 + \xi/2(n-1)) \qquad \ldots\ldots\ldots(2)$$

In Fig. 1 the effective node capacity is $5c/7$ and the network capacity is $15c/7$. From (1) and (2) the problem of maximizing capacity for given n and d clearly reduces to the forwarding index problem.

5.1. NOTATION AND TERMINOLOGY

More formally, let $G=(V, E)$ denote a network with vertex-set $V(G)$ and edge-set $E(G)$. If x,y are vertices in G, then a route is a path between x and y, denoted by $R(x, y)$. A routing R in G (graph G has n vertices) is defined as a set of $n(n-1)$ routes specified for all ordered pairs of vertices of G, one route for each ordered pair.

Let us call the *load of a vertex x* in a given routing R of a graph G, denoted by $\xi(G, R, x)$, the number of paths of R going through x (where x is not an end vertex). The vertex forwarding index of a network (G, R) is the maximum number of paths of R going through any vertex x in G and is denoted by $\xi(G, R)$,

$$\xi(G, R) = \max_{x \in V(G)} \xi(G, R, x)$$

The minimum forwarding index over all possible routings of a graph G will be denoted by ξ (G) and be called the *vertex-forwarding index of* G. The minimum taken over all the routings of shortest paths will be denoted by ξ_m (G).

$$\xi(G) = \min_{R} \xi(G, R) \quad \text{and} \quad \xi_m(G) = \min_{R_m} \xi(G, R_m)$$

Since the notion of load in networks (always limited in practice by the capacity) is at least as important for links as for nodes, it is interesting to introduce and study the same concepts for the edges of a graph.

Therefore we define the *load of an edge e* in a given routing R of G as the number of paths of R, which go through it, and denote it by $\pi(G, R, e)$. Then the edge forwarding index of (G, R), denoted by $\pi(G, R)$, is the maximum number of paths of R going through any edge of G

$$\pi(G, R) = \max_{e \in E(G)} \pi(G, R, e)$$

and the edge-forwarding index of G is defined as

$$\pi(G) = \min_{R} \pi(G, R) \quad \text{and} \quad \pi_m(G) = \min_{R_m} \pi(G, R_m)$$

Clearly $\xi(G) \le \xi_m(G)$ and $\pi(G) \le \pi_m(G)$. The equality however does not always hold as can be seen in the following example. The forwarding index of a cluster for shortest path routings is $O(n^2)$ and this is best possible.

Example: Let W_6 be the wheel on 7 vertices, with vertices 0,1,2,3,4,5 on a cycle and a vertex c joined to all the previous ones. Let us define routing of shortest paths R_m in W_6 as follows: for every i, $0 \le i \le 5$, $R_m(i, i+2) = R_m(i+2, i) = R_m(i, i+1, i+2)$ (where the vertices are taken modulo 6), and for $0 \le i \le 2$, $R_m(i, i+3) = R_m(i+3, i) = R_m(i, c, i+3)$. We have $\xi(W6, R_m, c) = 6$ and for any i, $0 \le i \le 5$, $\xi(W_6, R_m, i) = 2$, and clearly $\xi_m(W_6) = 6$. Also for any i, $\pi(W_6, R, ic) = 4$ and $\pi(W_6, R_m, i\ i+1) = 6$ and clearly $\pi_m(W_6) = 6$.

6. CONCLUDING REMARKS AND DETAILS FOR FURTHER STUDY

In this paper we have studied a very fundamental problem of how several nodes organize themselves into an ad-hoc network. We study heuristics for cluster formation and observe that the resultant work has the best possible forwarding index asymptotically. The randomized algorithm for cluster formation can be slightly altered to yield very good forwarding index. This algorithm has many applications, the foremost is to scatternet Bluetooth. According to Bluetooth specification, the smallest network unit is a piconet, consisting of a device and several Slave Bluetooth devices.

Bluetooth devices 'discover' each other by executing the Inquiry and Page procedures. In the Randomized algorithm, continuous Inquiry and inquiry scan is used. It is clear that an inquiry procedure can take a fair amount of time even for two devices to discover each other. Once a connection is established, any amount of information can be exchanged between the nodes without much overhead. The proposed study of forwarding index is very useful for load balancing by means of minimizing the load of the clusterhead (Master) node and this maximizes the life time of the clusterhead node.

The proposed study in our paper can be further extend to explore the broadcasting properties i.e. broadcasting radius. For good quality of clustering algorithms the low broadcasting radius can be one of the important parameter.

REFERENCES

[1]   Turgut D, Turgut B., Das S. K., Elmasri R,(2003), "Balancing Loads in mobile adhoc networks, Telecommunications, ICT 2003, 10th International Conference on, Volume:1, 23 Feb.1-March 2003.

[2]   F.Chung, E.Coffman, M. Reiman, and B. Simon, (1987), "The Forwarding Index of communication network," IEEE Transaction Info. Theory, 33(1987), pp. 224-232.

[3]   R. Saad, (1993), "Complexity of the Forwarding Index problem," LRI technical Report & SIAMS, Disc. Math., 6 (3), pp. 418-427, to appear.

[4]   A. Shastri, (1998), "Forwarding index and connectivity of Communication Networks," Proceedings of IEEE International Conference on Networking Indian and the World, Ahmedabad, pp 71-77.

[5]   C-C Chiang, (1997), "Routing in clustered multihop, mobile wireless network with fading channel," Proc. IEEE SICON' 97, pp. 197-211.

[6]   M. gary, Micheal R., and Johnson, David S.; Freeman, (1979), *"Computers and Interactablity:" A Guide to he theory of NP-ompleteness* NewYork.

[7]   Y. Manoussakis and Zs. Tuza;, (1989), "The Forwarding Index of directed networks," technical report L.R.I. No. 482, University of Paris-XI; Disc. Appl. Mathematics, 68, 1996, pp. 279-291. to appear.

[8]   Y. Manoussakis and Zs. Tuza;, "Optimal routings in communication networks with linearly bounded forwarding index," networks, to appear.

[9]   A. Aggarwal et. al., (2001), "Clustering Algorithms for Wireless Ad-hoc Networks,", preprint.

[10]   A. Shastri, (2001), "Wireless Ad-Hoc Networks with good Broadcasting and Load Balancing Properties," ATM Interact Symp. On Broadband Networking in the New millennium, New Delhi, India, pp. 43-50.

[11]   S.K. Das, S.K. Sen and R. Jayaram, (1998), "A Novel load Balancing   Scheme for the Tele-Traffic Hot speed Problem in Cellular Networks", ACM/Baltzer Journal on Wireless Networks, Vol. 4, No. 4, pp. 325-340.

[12]   S.K. Das, S.K. Sen and R. Jayaram, and P. Agarwal,(1997), "A distributed Load Balancing Algorithm for the Hot cell Problem in Cellular Mobile Networks", Proc. of Sixth IEEE International Symposium on High Performance Distributed Computing, Portland, Oregon, pp 254-263.

[13]   R. Prakash, A. D. Amis, (2000), "Load Balancing in Wireless ad-hoc networks", Application Specific Systems and Software Engineering Technology, Proc. 3rd IEEE Symphosium, 24-25.

Authors

Prof. Aditya Shastri is presently Professor of Computer Science & Vice Chancellor at Banasthali University. He received his M.Sc. (Tech.) Computer Science & M.Sc. (Hons.) Maths degree from BITS-Pilani, India before completing his Ph.D. at Massachusetts Institute of Technology, Cambridge USA in 1990. His research interests are in Discrete Mathematics, Graph Theory and Mobile Computing. He has written more than 50 research papers in journals of repute and authored 5 textbooks. Of late, he is engaged in the development of Banasthali University as its Chief Executive Officer and Chief Academic Officer.

Dr. Reena Dadhich is presently working as a Associate Professor and Head of the Department of Master of Computer Applications at Engineering College Ajmer, India. She received her Ph.D. (Computer Sc.) and M.Sc. (Computer Sc.) degree from Banasthali University. Her research interests are Algorithm Analysis & Design and Wireless Ad-Hoc Networks. She has more than 11 years of teaching experience. She has written many research papers.

ENERGY CONSERVATION IN MANET USING VARIABLE RANGE LOCATION AIDED ROUTING PROTOCOL

Nivedita N. Joshi[1] and Radhika D. Joshi[2]

[1]Department of Electronics and Tele-communication Engineering, College of Engineering, Pune, India
niv.joshi@gmail.com

[2]Department of Electronics and Tele-communication Engineering, College of Engineering, Pune, India
rdj.extc@coep.ac.in

ABSTRACT

A Mobile Ad-Hoc Network (MANET) is a temporary, infrastructure-less and distributed network having mobile nodes. MANET has limited resources like bandwidth and energy. Due to limited battery power nodes die out early and affect the network lifetime. To make network energy efficient, we have modified position based Location Aided Routing (LAR1) for energy conservation in MANET. The proposed protocol is known as Variable Range Energy aware Location Aided Routing (ELAR1-VAR). The proposed scheme controls the transmission power of a node according to the distance between the nodes. It also includes energy information on route request packet and selects the energy efficient path to route data packets. The comparative analysis of proposed scheme and LAR1 is done by using the QualNet simulator. ELAR1-VAR protocol improves the network lifetime by reducing energy consumption by 20% for dense and mobile network while maintaining the packet delivery ratio above 90%.

KEYWORDS

MANET, LAR1, Packet Delivery Ratio, Energy Consumption

1. INTRODUCTION

Wireless communication is one of the fastest growing fields in Telecommunication Industry. These systems, such as cellular, cordless phones, wireless local area networks (WLAN) have become an essential tool for people in today's life. Using these systems and the equipments like PDA, laptops, cell phones user can access all the required information whenever and wherever needed. All these systems need some fixed infrastructure. It takes time and potentially high cost to set up the necessary infrastructure. There are situations where user required networking connections are not available in a given geographic area, and providing the needed connectivity and network services in these situations becomes a real challenge. So, in this situation mobile communication network without a pre-exist network infrastructure is the best solution. The Ad Hoc Networks are wireless networks characterized by the absence of fixed infrastructures. The main aim of mobile ad- hoc network (MANET) [1] is to support robust and efficient operation in mobile wireless networks.

MANET consists of mobile nodes which form a spontaneous network without a need of fixed infrastructure. It is an autonomous system in which mobile hosts connected by wireless links are free to move randomly and often act as routers at the same time. Hence, it forms multi-hop network. The ad-hoc networks are finding more importance likely due to the features that they can be easily deployed as well as reconfigured. This allows the use of this kind of network in special circumstances, such as disastrous events, the reduction or elimination of the wiring costs and the exchange of information among users independently from the environment. The applications for MANETs are ranging from large-scale, mobile, highly dynamic networks, to small, static networks that are constrained by power sources [2]. It can be used in military communication, commercial sectors like disaster management, emergency operations, wireless sensor networks, etc.

Deployment of ad-hoc network leads to many challenges such as limited battery power, limited bandwidth, multi hop routing, dynamic topology, security [1]. But, the major issue in MANET is energy consumption since nodes are usually mobile and battery-operated. Power failure of a mobile node affects its functionality thus the overall network lifetime. To prolong life time of the network, ad-hoc routing protocol should consider energy consumption. Efficient minimum energy routing schemes can greatly reduce energy consumption and extends the lifetime of the networks. Different routing protocols are designed for MANET such as AODV, DSR, OLSR, ZRP, and LAR, which meets some of challenges explained above. But, they do not consider the energy-efficient routing. Here, we are proposing a scheme which uses energy efficient routing to route the data. The modification is done in position based LAR1 protocol.

Rest of the paper is organized as: a Section II different routing approach for MANET is discussed. Section III gives the brief overview of LAR1 protocol. The energy efficient routing schemes are surveyed in Section IV. Section V discusses a proposed idea. Section VI gives the detail description of simulation model and performance metrics used for experimentation. Comparative analysis of proposed and original LAR1 protocols using QualNet simulator is given in section VII. Finally, Section VIII concludes the paper with future work in Section IX.

2. DIFFERENT ROUTING APPROACHES FOR MANET

Many routing protocols have been developed for MANET with features like distributed operation, creation of loop free paths, security, and QoS support. MANET routing protocol is a standard that controls how nodes select the short and optimized route to route packets between communicating devices in a MANET. Routing protocols in MANET are broadly classified into proactive (Table-driven) and reactive (On-demand) routing protocols. These are based on routing strategy in MANET. Depending upon the network structure, the routing protocols are categorized as flat routing; hierarchical routing and geographic position assisted routing [3].

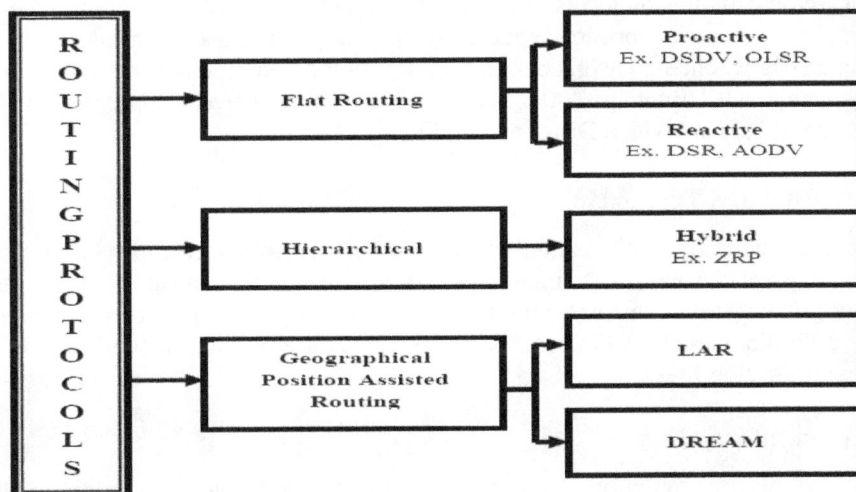

Figure 1. Classification of Routing Protocols

2.1. Proactive routing protocols

Proactive routing protocols maintain information in a table on each node about the routing to the other node in the network. Although the topology of the network does not change, this information must be updated periodically. Many proactive routing protocols have been proposed, for e.g. Destination Sequence Distance Vector (DSDV), Optimized Linked State Routing (OLSR) and so on.

2.2. Reactive protocols

These protocols don't maintain routing information or routing activity at the network nodes if there is no communication. If a node wants to send a packet to another node then this protocol searches for the route in an on-demand manner and establishes the connection in order to transmit and receive the packet. The route discovery usually occurs by flooding the route request packets throughout the network. Several reactive protocols have been proposed such as Dynamic Source Routing protocol (DSR), Ad hoc On-demand Distance Vector (AODV), and so on.

2.3. Hybrid routing protocols

For larger networks, it is complex to design routing protocols. All nodes in the network are separated into groups, called cluster. All clusters form a hierarchical infrastructure. For such networks separate routing protocols have been designed which are known as hybrid routing protocols. These are the combination of proactive and reactive routing protocols. They are used in large networks in order to take advantages of proactive scheme for maintaining the routes in a cluster and reactive scheme for retaining the routes between the clusters. Several hybrid routing protocols have been proposed such as Zone Routing Protocol (ZRP), Zone-based Hierarchical Link State (ZHLS).

2.4. Geographic position based Routing protocols

The above mentioned types of MANET routing protocols do not consider the geographical location of a destination node. Position-based routing protocols do not establish or maintain route, a packet is forwarded one hop closer to its final destination by comparing the location of destination with the location of the node currently holding the packet. These routing protocols make minimum use of the topology information, hence, they exhibit better scalability compared to topology-based routing protocols. Such protocols use 'localized' algorithms like greedy forwarding, in which a node forwards a packet to a next hop that is geographically closest to the destination among its one-hop neighbors. Several Position-based protocols have been proposed such as Location aided Routing (LAR) [4], Distance Routing Effect Algorithm for Mobility (DREAM), Most Forward within Distance R (MFR).

3. INTRODUCTION TO LAR1

Position-based routing protocols exhibit better scalability, performance and robustness against frequent topological changes. These routing protocols use the geographical location of nodes to make routing in networks. This will improve efficiency and performance of the network. The main aim of Position-based LAR1 described in [4] is to reduce the control overheads by the use of location information.

3.1. Location Information

LAR1 protocol requires the information about geographical location of the nodes in network. This location information can be determined by using Global Positioning System (GPS) [5]. By using location information, LAR1protocol limits the search for a new route to a smaller request zone of the ad hoc network. This results in a significant reduction in the number of routing messages.

3.2. Estimation of Expected and Request Zones

LAR1 has two types of zone, Expected zone and Request zone, to restrict the flooding of route request packets. A source node uses the location service to find out the location of the destination and according to that information it will set the expected zone. Request zone is also determined by the source node and it is zone where a route request should be forwarded from source.

3.2.1. Expected Zone

Expected zone is set up by the source node S when it has data intended for destination node D. By using location service node S estimates the geographical location of node D at time t_0. Suppose node D was at location O at time t_0, and that the current time is t_1. From this information node S is capable of determining the 'expected zone' of node D from the viewpoint of node S by time t_1. It is the region that node S expects to contain node D at time t_1. For instance, if node D is travelling with average speed v, then node S assumes that node D is in the expected zone of circular region of radius $v(t_1 - t_0)$, centered at location O. The expected zone is only an estimate made by node S to determine a region that may contain D at time t_1. Since, if actual speed of node D is greater than the average, then the destination D may actually be outside the expected zone at time t_1. Figure 1 shows the expected zone created by the source node S.

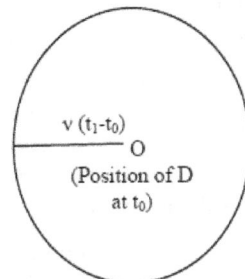

$v(t_1-t_0)$

O
(Position of D
at t_0)

Figure 2. Expected Zone

3.2.2. Request Zone

The 'request zone' is different from the expected zone. It is the zone where a route request should be forwarded from source. An intermediate node will forward a route request packet only, if it belongs to the request zone. The request zone should contain the expected zone to reach destination node D. The source node S defines this zone for flooding the route request packets. An intermediate node will forward the request packet, only if it is located within the request zone.

3.3. Protocol Functioning

In LAR1 scheme the request zone is of a rectangular shape. Assumption is that the source node S knows the destination node D's average speed v and location (X_d, Y_d) at time t_0. The request zone is considered to be the smallest rectangle that includes the current source location, and the expected zone. The sides of the rectangle are parallel to X and Y axes. The source node determines the request zone and initiates the route request packet containing the four corners coordinates of request zone. Each intermediate node, receiving the request packet, checks whether it belongs to the rectangle; if it does not, it will discard the packet such as node Q does in the following Figure 2. In the same Figure 2, node P will forward the packet because it belongs to the rectangle. In the reply packet, node D will attach its accurate location and the current time stamp, which will be stored in the source's cache for future use.

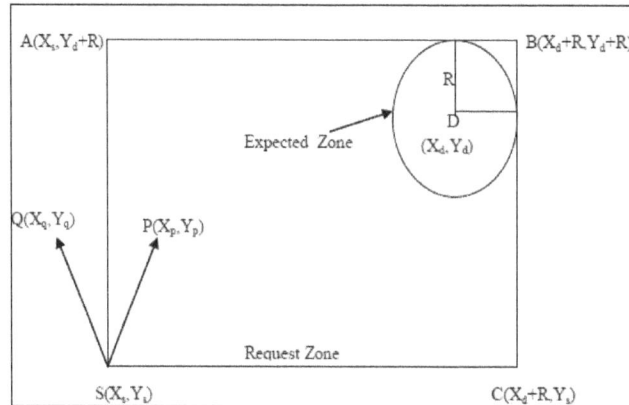

Figure 3. Working of LAR1

4. BRIEF OVERVIEW OF ENERGY EFFICIENT ROUTING SCHEMES

Several energy-efficient techniques are proposed to reduce energy consumption in MANET. These techniques use energy aware metrics to establish a path in a network. These metrics are residual energy, transmission power or link distance.

Niranjan Kumar Ray and Ashok Kumar Turuk have discussed different energy efficient techniques for wireless ad-hoc network [6]. One of the techniques is based on reduction of number of route request messages. In second Power control technique, next hop node is chosen depending on the power level of the node. Topology control technique is used to remove the energy-inefficient link from the network by examining the power level of the node. This technique helps network devices to take decision about their transmission range.

Morteza Maleki, Karthik Dantu, and Massoud Pedram in [7] have proposed a new power-aware source-initiated (on demand) routing protocol for mobile Ad-hoc networks that increase the network lifetime up to 30%. A greedy policy was applied to fetched paths from the cache to make sure no path would be overused and also make sure that each selected path has minimum battery cost among all possible path between two nodes. Power-aware Source Routing (PSR) has taken care of both the node mobility and the node energy depletion that may cause a path to become invalid.

Energy-based Route Discovery Mechanism in Mobile Ad Hoc Networks [8] selects the route which has lowest energy cost in the network. The energy cost represents energy consumption of the network in order to prolong all connections between source and destination nodes. The energy cost is calculated using realistic energy consumption modeling which is used the channel quality to decide whether each packet is successfully received.

Location aided Energy-Efficient Routing protocol (LEER) protocol finds out the all possible paths from source to destination and selects minimum energy path to route the packets [9]. The selection of next hop node is based on whether it is situated near to destination than to source as well as transmit power of that node.

An energy aware routing scheme in location based ad-hoc network has proposed by Jangsu Lee, Seunghwan Yoo and Sungchun KimIn [10]. This method modifies the LAR protocol in which the virtual grid is applied to ad hoc network region and high energy node is selected as header for each grid which communicates information about nodes in that particular grid. The transmit power of nodes is adjusted according to the distance between them. The next hop node will be selected based on transmit power and its distance from the destination.

Nen-Chung Wang and Si-Ming Wang [11] have proposed a scheme which decides the baseline line between the source node and the destination node, for route discovery. The next hop is then selected based on baseline by broadcasting the request packets in request zone. The neighboring node with the shortest distance to the baseline is chosen as the next hop node. This method reduces control overheads by finding a better routing path than LAR scheme. They have proposed a partial reconstruction process for maintaining broken links of routing path.

Arthi Kothandaraman has proposed a protocol which based on transmission power control [12]. It varies the transmission range of a node to exclusively accommodate an independent node's neighbor set. This purely distributed as well as protocol independent scheme and preserves connectivity, and allows low power transmissions.

5. PROPOSED SCHEME

LAR1 protocol uses location information of a node for setting the path from source to destination. We take this feature of LAR1 as a key factor in designing of variable range technique. The main aim is to design a technique of variable transmission power control to reduce overall energy consumption of the network. RREQ in LAR1 protocol consists of source location and destination location information. We have used this information to calculate the distance between the nodes. We also embed the energy factor of the node in RREQ packet for selection of energy efficient path.

5.1. Energy Factor Calculation

Yang Qin et. al. [13] have proposed an energy efficient routing metric called as Energy factor. They have used this metric for multipath concept where the most energy efficient as well as shortest path is selected to deliver the data packets. Similarly, in our method we use this metric to select the next hop node while discovering the path to destination. We define an energy-efficient routing metric for selecting the node having sufficient energy to route the packets. The terms used in metric are as follows:

EF_p: Energy Factor of node p

RE_p: Remaining Energy of node p

IE_p: Initial Energy of node p

Energy factor is calculated by using the formula (1) for every node when it receives route request or data packet.

$$EFp = REp \ / \ IEp \qquad (1)$$

The EF of all the nodes along a valid path are multiplied together to obtain the EF of the path. We define it as minimum cost and it is given by-

$$Minimum\ cost = \prod_{p \in N} EFp \qquad (2)$$

Where, N: Number of nodes between source S and Destination D.

Minimum cost metric selects the most energy-efficient path. The purpose of multiplication of EF values is to select the nodes having sufficient energy so that the minimum cost of the path from source to destination is high.

5.2. Flow Chart for ELAR1-VAR

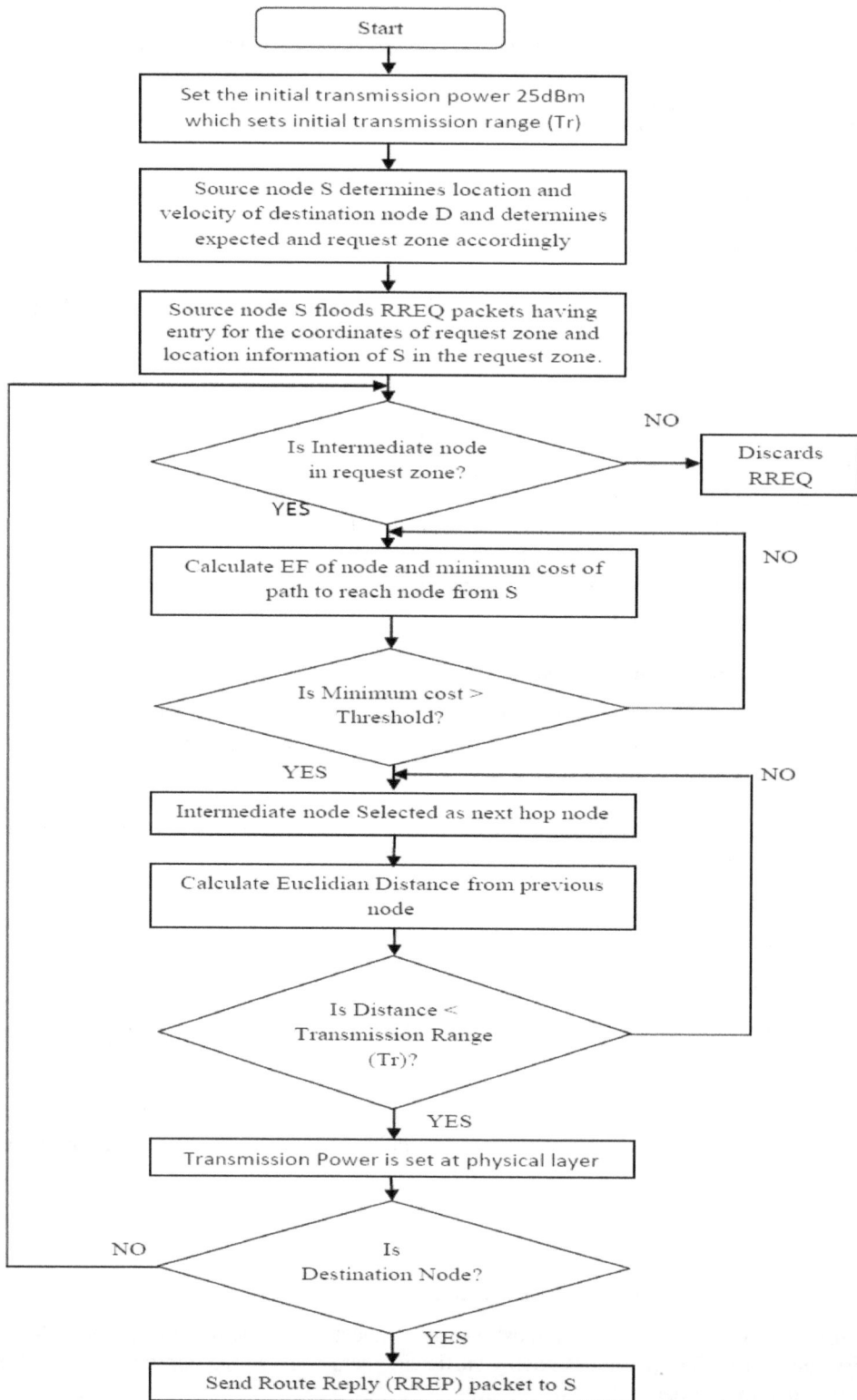

Figure 4. Flow Chart for ELAR-VAR

6. SIMULATION MODEL AND PERFORMANCE METRICS

Simulator used for the comparative analysis is QualNet 5.0 [14]. It is suitable for designing new protocols, comparing different protocols and traffic evaluations. QualNet architecture is divided into three levels, Kernel, Model Libraries and Graphical User Interface (GUI).

1. Kernel- It provides the scalability and portability to run hundreds of nodes.

2. Model Libraries- Different types of model libraries enable to design networks using protocol models.

3. QualNet GUI- The user can use the QualNet GUI for creating and animating network scenarios, and analyzing their performance using the analyzer.

6.1. Important features of QualNet –

1. Speed- QualNet enables designers to run multiple scenarios by varying model, network, and traffic parameters in a short time.

2. Scalability- QualNet can model thousands of nodes by taking advantage of the latest hardware and parallel computing techniques.

3. Model Fidelity- QualNet uses highly detailed IEEE standards-based implementation of protocol models.

4. Portability- QualNet and its library of models run on a vast array of platforms, including Windows XP, Mac OS X, and Linux operating systems

5. Extensibility- QualNet can connect to other hardware and software applications, such as OTB, real networks to greatly enhancing the value of the network model.

6.2. Simulation Model

Simulation model is comprised of traffic model, mobility model, energy model, and battery model. The detailed description of all these are as follows:

6.2.1. Traffic Model and Mobility Model

Traffic is generated using constant bit rate (CBR). The packet size is limited to 512 bytes. The mobility model is used to describe the movement pattern of mobile users, and how their location, velocity and acceleration change over time. Random waypoint mobility model is used to give mobility to nodes in the network.

6.2.2. Energy Model

Energy model is used to analyze the overall energy consumption of the network. QualNet supports generic energy model [15]. We have used generic energy model for experimentation because it can estimate energy consumption in transmitter for the case of continuous and variable transmit power levels. Table 1, shows specifications of generic energy model.

Table 1. Generic Energy Model

Parameter	Values
Power Amplifier Inefficiency factor	6.5
Transmit Circuitry power Consumption	100.0 mW
Receive Circuitry power Consumption	130.0 mW
Idle Circuitry power Consumption	120.0 mW
Sleep Circuitry power Consumption	0.0
Supply voltage (volt)	6.5

6.2.3. Battery Model

MANET consists of mobile nodes which are battery operated. QualNet has Linear Battery model which enables the analysis of the discharge behavior of the battery in mobile nodes.

6.3. Performance Metrics

Quality of a service is the performance level of a service offered by the network. The multimedia services demand for quick and better quality of data, it means the data rate and delay are the key factors for these services. For military applications, the security and reliable data delivery are of major concern. In sensor networks, the energy consumption and battery life would be the most important parameters. Different performance metrics are used for comparison are as given below-

6.3.1. Packet Delivery Ratio

It is the ratio of total number of data packets received successfully at destination to number of data packets generated at the source [16]. PDR values range from 0 to 1. Higher PDR values decide the consistency of the protocol.

6.3.2. End-to-End Delay

The end to end delay is the average time interval between the generation of a packet at a source node and the successfully delivery of the packet at the destination node [16]. Low end to end delay gives better performance of the network.

6.3.3. ECSDD (Energy Consumption per Successful Data Delivery)

It is the ratio of total energy consumption to the number of data packets successfully delivered to the destination [17]. Lower the ECSDD values indicate that node uses less energy for data communication. This helps in extending the lifetime of node and thus overall network lifetime.

6.3.4. Average energy consumption

Energy consumed by all nodes in the network [17]. Energy consumption should be as low as possible so as to make fair utilization of limited battery power.

7. SIMULATION RESULTS AND DISCUSSIONS

The network area taken for experimentation is 1000m x 1000m. The comparison of proposed and original protocol is done for different number of nodes, speeds, packet rates and number of connections (Traffic load). Battery model is used at the node level with battery capacity 60mAHr .We have set the transmission power for nodes equal to 25 dBm. The scenario model for examining the ELAR1-VAR in dense and moderately mobile network is given in Table 2.

Table 2. Scenario Model

Parameters	Variation of Number of Nodes	Variation of Node speed	Variation of Packet rate	Variation of Number of connection
Area(m^2)	1000 x 1000	1000 x 1000	1000 x 1000	1000 x 1000
Nodes	20,40,60,80,100	50	50	50
Node	10	1,5,10,15,20,25,30	10	10
Simulation	900	900	900	900
Pause Time(s)	100	100	100	100
Traffic Type	CBR	CBR	CBR	CBR
No. of	25	25	25	25,30,35,40,45,50
Packet	4	4	2,4,6,8,10	4
Initial Power	60 mAHr	60 mAHr	60 mAHr	60 mAHr

7.1. Impact of Variation of Number of Nodes

Simulation parameters for analysis of variation of number of nodes are given in Table 2. Both protocols are analyzed for variation of nodes from 20 to 100. This will help in analyzing the behavior of protocols in dense and sparse network.

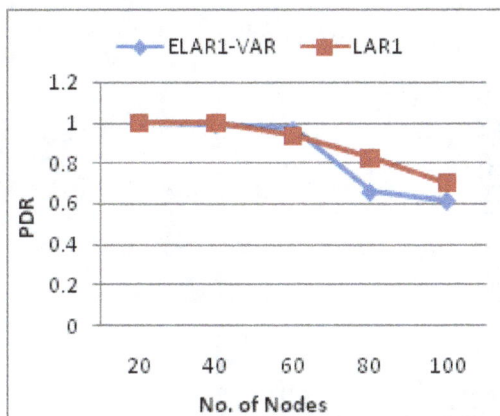

Figure 5. PDR vs. No. of Nodes

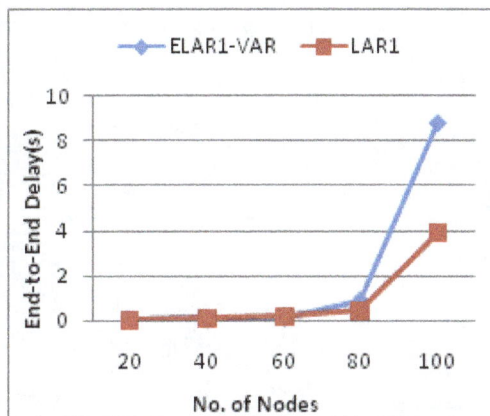

Figure 6. End-to-End Delay vs. No. of Nodes

The variation of PDR and end to end delay with respect to number of nodes is shown in Figure 5 and Figure 6 respectively. It is observed that ELAR1-VAR has same packet delivery as LAR1 up to 60 nodes. PDR for highly dense network is slightly less. We can say that ELAR1-VAR is suitable option for low as well as dense network. End to end delay for modified protocol is equal to original one but for highly dense network it increases rapidly. This is because the modified method calculates the distances between the nodes for setting transmit power which increases the processing time of RREQ at intermediate node.

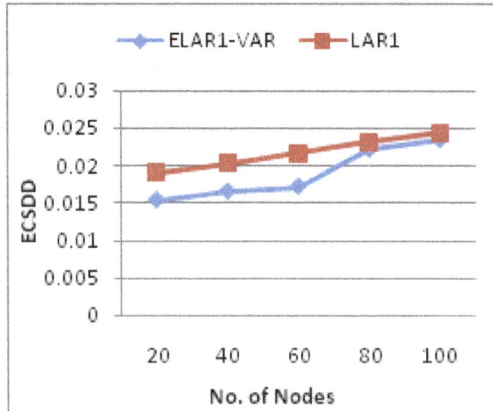

Figure 7. ECSDD vs. No. of Nodes

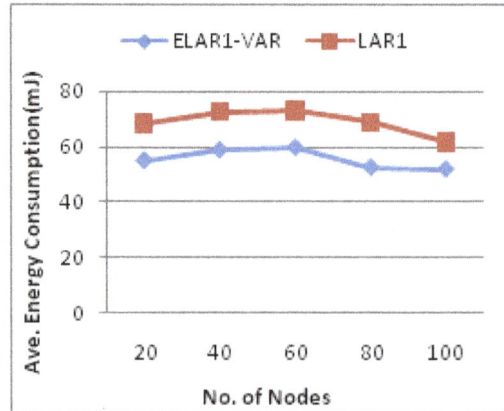

Figure 8. Ave. Energy Consumption vs. No. of Nodes

Figure 7 and Figure 8 shows ECSDD graph and energy consumed for different number of nodes respectively. The modified protocol consumes low energy for each successful data delivery. Hence, we can use ELAR1-VAR protocol for energy efficient applications. It is clear from figure 8 that energy consumption for ELAR1-VAR decreases approximately by 20%.

7.2. Impact of Variation of Node Speed

We are varying the speed of mobile nodes from 1m/s to 30 m/s to see the performance of ELAR1-VAR in mobile environment. The results for different performance metrics are as follows:

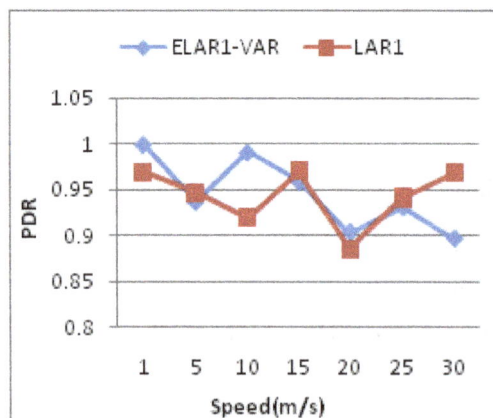

Figure 9. PDR vs. Speed

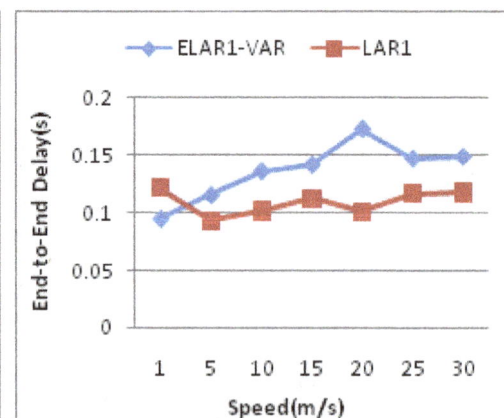

Figure 10. End-to-End Delay vs. Speed

The PDR and end to end delay results for different speeds are as shown in Figure 9 and Figure 10 respectively. If we increase the speed from 1 to 30 then PDR for ELAR1-VAR is approximately above 90% which indicates that it gives better results in low as well as high mobility network. There is an increase in delay for modified protocol. This is because location of node changes rapidly as node speed increases and node has to send new control packets which results in more delay.

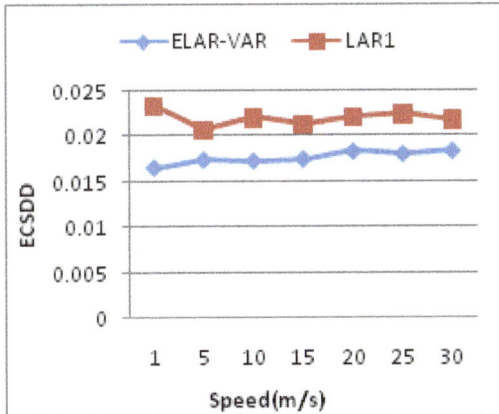

Figure 11. ECSDD vs. Speed

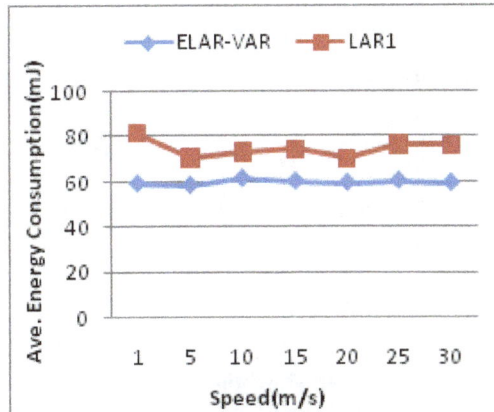

Figure 12. Average Energy Consumption vs. Speed

Impact of speed variation on ECSDD is shown in Figure 11. Energy consumption and ECSDD for ELAR1-VAR is less than LAR1. But, as speed increases the energy consumption in Figure 12 for both the protocol remains same. It is clear from Figure 11 that the energy consumption per data delivery decreases which improves the battery lifetime.

7.3. Impact of Variation of Packet Rate

Packet rate is varied from 2 to 10packets/second. Packet rate is varied to analyze the network capability of handling large number of data at a time.

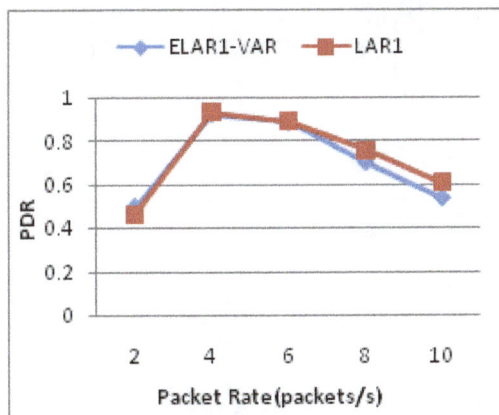

Figure 13. PDR vs. Packet Rate

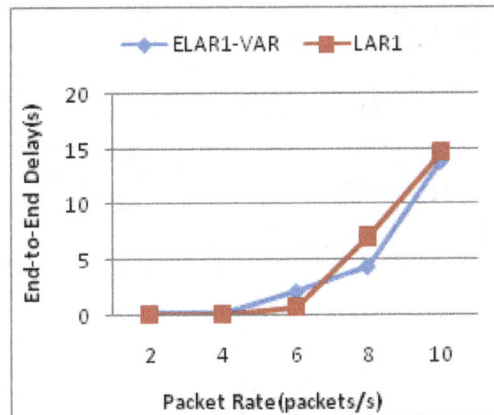

Figure 14. End-to-End Delay vs. Packet Rate

Figure 13 gives performance comparison of ELAR1-VAR with LAR1 in terms of PDR. There is increasing trend of PDR for both ELAR1-VAR and LAR1. As shown in Figure 14, ELAR1-VAR has comparable delay value for lower packet rates. Delay increases with the packet rate for both protocols. Since, the more packets are to be sent at a time which overflows the buffer resulting in dropping of packets. Hence, packets are to be retransmitted which increases the end to end delay.

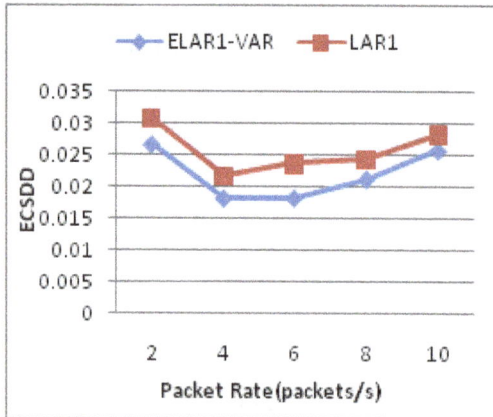

Figure 15. ECSDD vs. Packet Rate

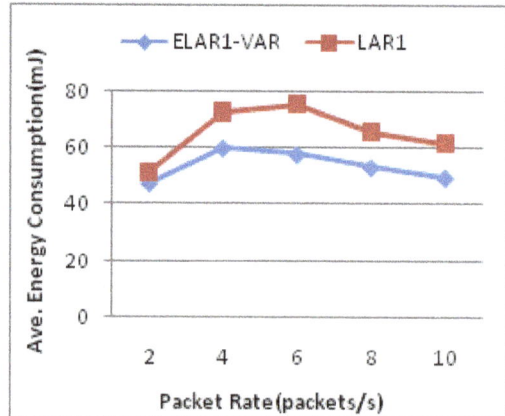

Figure 16. Average Energy Consumption
vs. Packet Rate

As depicted in Figure 15, there is noteworthy change in ECSDD value for modified protocol. Also as shown in Figure 16, the overall energy consumption for network decreases with ELAR1-VAR.

7.4. Impact of Variation of Number of Connections

The number of connections i.e. source to destination data transfer connections are varied from 25 to 50 in steps of 5. This evaluation helps in understanding the behavior of network under variable traffic load.

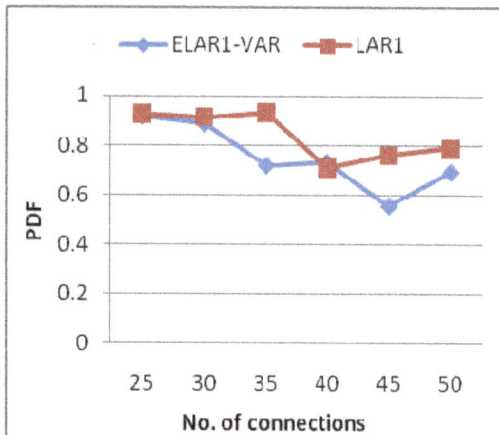

Figure 17. PDR vs. No. of connections

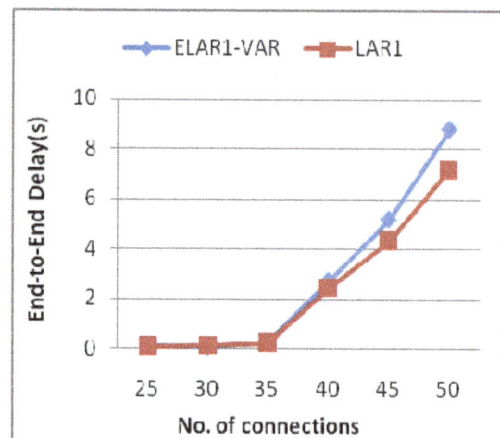

Figure 18. End-to-End Delay vs. No. of
connections

Variation of PDR against number of connections is given in Figure 17. PDR decreases for both the protocols as traffic load increases. This happens due to the increase in traffic load creates the congestion in the network which will result into packets drop. End-to-end delay given in Figure 18 for LAR1 and ELAR-VAR1 is having same value up to 35 connections. For higher traffic loads; there is increasing trend for both the protocols. For the higher traffic loads, the queue at each node will be full and creating the congestion in the network. Thus, data delivery is delayed.

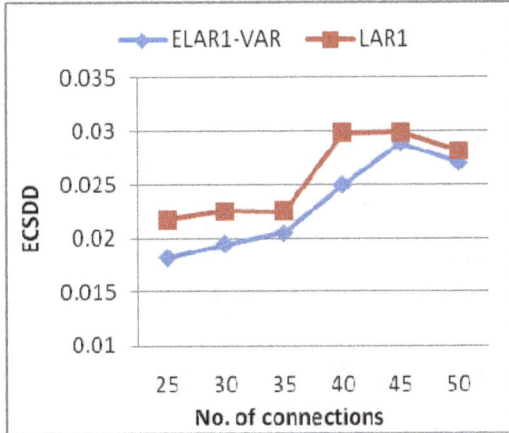

Figure 19. ECSDD vs. No. of connections

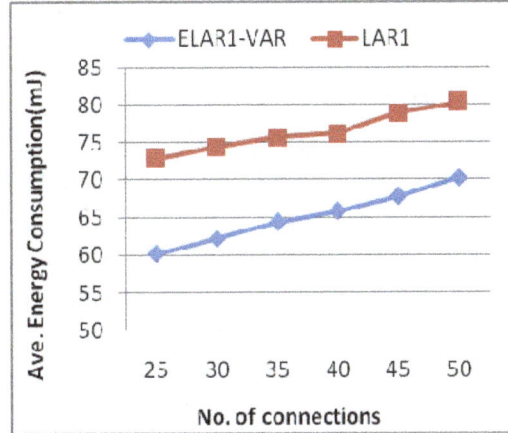

Figure 20. Average Energy Consumption vs. No. of connections

ECSDD and Average energy consumption graphs are given in Figure 19 and 20 respectively. Energy consumption increases linearly with the traffic load. For entire range of number of connections for ELAR1-VAR energy consumption is less as compared to LAR1. ECSDD values for the both protocols are remained almost constant up to 35 CBR connections after that it increases with traffic load. ECSDD values are decreasing after 45 connections due increase in energy consumption and less packet delivery.

8. CONCLUSION

We compared ELAR1-VAR and LAR1 protocols considering the performance metrics such as packet delivery ratio, end to end delay, average energy consumption, and ECSDD. Several simulations have performed under different network conditions to analyze the performance of modified protocol with LAR1. The analysis of results showed that there is very slight change in packet delivery but remarkable change in energy consumption. Packet delivery is low for highly dense network and for large traffic loads whereas for all speed variation it maintains 90% of PDR. Average end to end delay is less sensitive to node speed. Lower packet rate and traffic load have less impact on delay. It increases rapidly for higher packet rate and highly dense network.

The ELAR1-VAR consumes less energy for dense and moderately mobile network having packet rate of 4 packets/second. Network performance is better for traffic load about 50% of nodes i.e. for 25 CBR connections. It is observed that overall energy consumption of the network is decreased by 20%. In ELAR1-VAR, the energy consumption per successful data delivery is lowered and thus nodes in the network can communicate with each other for longer period. Thus, our aim of extending network lifetime is achieved. ELAR1-VAR due to its energy efficient feature can be used for the energy constrained applications.

9. FUTURE WORK

The work can be extended to analyze the performance of ELAR1-VAR for different performance metrics such control overheads, jitter and throughput.

In future, we will try to embed technique which will set threshold on the basis of nodes drain rate. This will help in selection of next hop node accurately.

REFERENCES

[1]   Imrich Chlamtac, Marco Conti & Jennifer J.-N. Liu, (2003) "Mobile ad hoc networking: imperatives and challenges", *Ad-hoc Networks, Elsevier*, pp13-64.

[2]   Jun-Zhao Sun, (2001) "Mobile Ad Hoc Networking: An Essential Technology for Pervasive Computing", *International Conference on Info-tech and Info-net*, Vol. 3, pp316-321.

[3]   Krishna Gorantala, (2006) "Routing Protocols in Mobile Ad-hoc Networks", Master Thesis in Computer Science, pp1-36.

[4]   Young-Bae Ko & Nitin H. Vaidya, (1998) "Location-Aided Routing (LAR) in Mobile Ad Hoc Networks", *MOBICOM 98*, pp66-75.

[5]   G. Dommety & R. Jain, (1996) "Potential networking applications of global positioning systems (GPS)", Tech. Rep. TR-24, CS Dep., The Ohio State University.

[6]   Niranjan Kumar Ray & Ashok Kumar Turuk, (2010) "Energy Efficient Techniques for Wireless Ad Hoc Network", *International Joint Conference on Information and Communication Technology*, pp105-111.

[7]   Morteza Maleki, Karthik Dantu & Massoud Pedram, (2002) "Power-aware Source Routing Protocol for Mobile Ad Hoc Networks", *Iternational Symposium on Low Power Electronics and Design (ISLPED)*, pp72-75.

[8]   J. Kanjanarot, K. Sitthi, and C & Saivichit, (2006) "Energy-based Route Discovery Mechanism in Mobile Ad Hoc Networks", ICA0T2006, pp1967-1972.

[9]   Dahai Du & Huagang Xiong, (2010) "A Location aided Energy-Efficient Routing Protocol for Ad hoc Networks", *19th Annual Wireless and Optical Communications Conference (WOCC)*, pp1-5.

[10] Jangsu Lee, Seunghwan Yoo & Sungchun Kim, (2010) "Energy aware Routing in Location based Ad-hoc Networks", *Proceedings of the 4th International Symposium on Communications, Control and Signal Processing (ISCCSP)*, pp3-5.

[11] Nen-Chung Wang & Si-Ming Wang, (2005) "An Efficient Location-Aided Routing Protocol for Mobile Ad Hoc Networks", *11th International Conference on Parallel and Distributed Systems (ICPADS'05)*, Vol. 1, pp335-341.

[12] Arthi Kothandaraman, (2008) "An Energy-Efficient Transmission Power Control Protocol for Cooperative Robotics", Master Thesis, Auburn University, Alabama, pp1-63.

[13] Yang Qin, Y.Y. Wen, H.Y. Ang & Choon Lim Gwee, (2007) "A Routing Protocol with Energy and Traffic Balance Awareness in Wireless Ad Hoc Networks", *6th International Conference on Information, Communications & Signal Processing*, pp1-5.

[14] QualNet details :http://www.scalablenetworks.com.

[15] T.V.P.Sundararajan, K. Rajesh Kumar & R.K. Karthikeyan, (2009) "A Novel Survey towards Various Energy Models with AD HOC on Demand Distance Vector Routing Protocol (AODV)", *International Conference On Control, Automation, Communication And Energy Conservation*,pp1-5.

[16] Dr. S. Karthik, S. Kannan, Dr. M.L. Valarmathi, Dr. V.P. Arunachalam & Dr. T. Ravichandran, (2010) "An performance Analysis and Comparison of multi-hop Wireless Ad-hoc Network Routing protocols in MANET", *International Journal of Academic Research*, Vol. 2, No. 4, pp119-124.

[17] Alemneh Adane, (2008) "Active Communication Energy Efficient Routing Protocol Of Mobile Ad-hoc Networks (MANETS)", A thesis submitted to the school of Graduate studies of Addis Ababa University, Ethiopia.

Authors

Nivedita N. Joshi received the BE degree in 2008 from Pune University. She is currently a M.Tech student at College of Engineering, Pune, India which is an autonomous institute of Government of Maharashtra. She is pursuing her M.Tech in **Wired and Wireless Communication.** Her area of interests are wireless communication, image processing.

Radhika D Joshi received the BE degree in 1993 from Pune University and ME Degree from Pune University in 2002. She is currently a PhD student at the University of Pune. She is pursuing research in the field of **Wireless Mobile Ad hoc networks energy management issues.** At present she is serving as Assistant Professor at Electronics and Telecommunication Engineering department of College of Engineering Pune, which is an autonomous institute of Government of Maharashtra. Her areas of interests are wireless communication, signal processing and electronic devices and circuits. She has received grants for two research proposals from two funding agencies recently. She has published several papers in journals and international conferences.

3

ENERGY-LATENCY IMPROVED SENSOR NETWORKS USING MOBILE AGENTS IN TEXTILE INDUSTRY

G.Sundari[1] and P.E.Sankaranarayanan[2]

Research scholar, Sathyabama University, Chennai, India
sundari.sundarc@rediffmail.com
Dean (Academic studies), Sathyabama University, Chennai, India
drpesanky37@hotmail.com

ABSTRACT

The key performance metrics in wireless sensor networks are both network lifetime and an average time required to report an event reliably. The optimal solution must be taken into account to meet these metrics. Considering energy-latency constraints in sensor networks, we have developed a cluster architecture using mobile agents to detect the yarn break in spinning machine in textile industry. We have used the latency improved MAC layer protocol to access the channel in the cluster and also used energy efficient cluster formation algorithm based on residual energy of the sensor node. This protocol can handle multiple event messages efficiently. And, we mainly focused on simulating the behavior of mobile agents for information retrieval from a wireless sensor network in a event driven and poll driven wireless sensor network model.

KEYWORDS

Latency, MAC protocol, Energy efficiency, Sensor node, Mobile agent

1. INTRODUCTION

Wireless sensor networks (WSN) are gaining popularity for industrial sensing applications due to their relatively low cost and simplicity for retrofitting into existing infrastructure. WSN is a network consisting of distributed self–organized autonomous devices built with sensors to monitor physical and environmental conditions such as vibration, motion, temperature and sound in a coordinated fashion. It is a group of specialized transducers with a communication infrastructure intended to monitor and record conditions at diverse locations[1]. They are particularly suited to power engineering applications as they do not require cabling, which will lead to shorter outages during installation and a lower capital outlay than their wired equivalent. In emerging Wireless Sensor Networks, including Sensor Webs, a 'deploy and ignore' approach is no longer possible[2]. Indeed, the conventional 'built-in' station is replaced by a large number of lower-power motes which often conserve their resources by having sleep/wake cycles. In addition, in mobile data collection, communication pathways and network topologies regularly change. Consequently, monitoring must be actively and continually updated. Especially in remote areas where manual maintenance is nearly impossible, monitoring must be carried out by the network in a way that it can detect events of interest and selfconfigure quickly and efficiently in order to collect and forward data to sinks.

Generally, there are three models of wireless sensor network: continuous, on-demand and event-driven. In continuous WSN, sensors send data periodically to the sink. There are always sensors in the network, which initiate the communication. In the on-demand WSN model, sensors send data only when they receive a request from a access point. Without request, sensors sense the information and store it in their local memory. In the event driven WSN, in which we are interested, the sensors send data only when an event occurs[3]. We made an attempt to use an mobile agent-based system that will provide the abstraction for the different technologies used in spinning machine. Here the agent has the ability to communicate with the sensors in the environment, learn about the status of all sensors, analyze the decisions, and acts autonomously. The agent acts as a mediator between the administrator and the sensor network environment. This agent provides real-time solutions for energy-efficient query processing and data dissemination. The agent-based architecture will use a database approach where the users will be unaware of the implementation details. In other words, users would be able to query the network without worrying about the energy efficient query processing and routing techniques. The query and routing agents would be responsible for query processing and data dissemination respectively. Moreover, the application developers could easily build a standard application for the given sensors, without going into the hardware, routing, or sensor details[4]. Figure 1 shows the client-server based and mobile agent based architecture. Shaded circles represents the nodes.

In this paper, section 2 states the problem description. Section 3 describes some related MAC protocols and the energy efficient clustering formation algorithm, Section 4 details the background of mobile agents and the proposed model. In section 5, we showed the simulation results and section 6 concludes the paper.

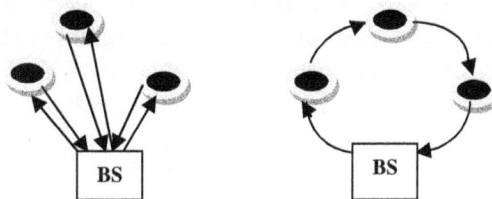

Fig 1a.Client-server based   Fig 1b. Mobile Agent based

2. PROBLEM DESCRIPTION

Detecting the yarn break in a spinning machine in a textile industry is the important task as it directly affects the production efficiency. The end breakage in ring spinning not only reduces the running efficiency of the process, but also deteriorates the quality of the yarn in terms of presence of piecing slubs. With the increase in spindle speed, the yarn breakage rate increases. A higher speed causes powerful centrifugal forces on the fibres resulting in poor spinning stability and an increase in yarn tension, which in turn causes high yarn breakage. An appropriate sensor network is always required to monitor the system behavior in such a way that suitable signal processing and damage-sensitive feature extraction on the measured data can be performed efficiently. The number of sensing systems available vary significantly depending upon the activity. Current sensing network in spinning machine uses wired concept. Wireless communication can remedy the cabling issues with the traditional monitoring system and significantly reduce the maintenance costs. A new wireless sensor network (WSN) paradigm, which has been investigated by the authors, is then presented.

3. MAC PROTOCOLS AND CLUSTERING ALGORITHM

Research on the medium access control of wireless sensor networks is very common. There is a need to improve MAC protocol management of communication time between sensors which consumes the most energy. MAC protocols are classified into two different types: Contention-Based and Contention-Free. Contention-free MAC is based on reservation and scheduling[5]. Here, each node announces a time slot that they want to use to the coordinator of the network. This coordinator schedules the request and allocates to other nodes their respective time slots. In this way, a node can access the channel without colliding with others because it is the only node which can transmit during its time slot.   This technique guarantees low energy consumption because each node in the network works only in its time slot without collisions. However, the main drawback of this technique is,the nodes must be well synchronized among them (about several μs), which is not easy to achieve in the widely distributed and scalable environment of a sensor network. Unlike this technique, contention-based MAC is a protocol where every node competes to access the channel. Before transmitting a message, a node listens to the channel to see whether there is already a transmission in the medium. If the channel is busy, it will wait for a random time and retry to detect it later. If the channel is free, it will transmit the message. Collision occurs when two or more interfering nodes observe that the channel is free at the same time and they transmit their message simultaneously[6]. The most well-known example of this technique is protocol IEEE 802.11 for wireless LAN network. Many research projects have been carried out to optimize the existing MAC methods. S-MAC[7] is considered to be the first MAC protocol proposal which tries to reduce idle time for sensors, it is not suitable for our event driven application. TMAC tries to improve SMAC by using a time out scheme to further reduce the idle listening wastage. DMAC[8] uses an improved staggered sleeping cycle to reduce the data latency of SMAC. Latency MAC protocol[9] sacrifices some packet latency to reduce the energy wastage of idle listening and to increase the accuracy of the prediction for packets incoming. Sift[10] is the MAC protocol specific to event-driven WSN and guarantees successful transmission of messages in the shortest delay but the distribution function is complicated and requires a lot of processing to compute the non-uniform probability for each transmission. So we have adapted Sift to improve the latency and to meet our event driven application.

Our work is motivated by the following observations: In most sensor networks, multiple sensors are deployed in the same geographic area, usually for fault-detection and reliability. In addition to sensing periodic observations, when an event of interest happens, the sensing node that senses the event send messages reporting the event. The result is spatially-correlated contention. Multiples sensors sharing the wireless medium all have messages to send at almost the same time because they all generate messages in response to external events. In our work, each node is deployed across each spindle to monitor the rotation of the traveler. So, only a single node is responsible for sending the event of interest.

In all contention window based protocols, each node picks a random contention slot in [1,CW] to transmit. If two nodes pick the same slot, they will both transmit and cause a collision. When this happens, most protocols specify that the colliding nodes multiplicatively increase their value of CW. In contract to this, sift protocol uses a small and fixed CW of 32 slots, where each slot lasts on the order of tens of microseconds. In a shared medium, where R nodes sense an event and contend to transmit on the channel at the same time, we design a MAC protocol that minimizes the time taken to send R messages without any collision. The key idea is to use an increasing non-uniform probability distribution with in a fixed contention window, rather than varying the contention window size as in many traditional MAC protocols.

If exactly one node happens to pick some contention slot, it will start to transmit in that slot. When its transmission finishes, all other competing nodes select new random contention slots and repeat the process of backing off over the fixed contention window. The same process happens if two or more stations happen to pick the same contention slot resulting in collision. So in our method, a node with larger report information is delayed by the other node.

3.1. Energy efficient cluster formation algorithm

Energy consumption of the cluster heads is relatively expensive, so the residual energy of sensor node is the main criteria for the election of cluster head[11][12]. The closer sensor nodes within a cluster, the lower energy they need to consume to send the data. Considering a sensor network which is randomly distributed over a region with a sink node. The sink node is located outside of the monitoring area with infinite energy. All these sensor nodes are grouped into clusters. All sensor nodes are assigned with an ID(identification), computing power and processing power. Initially some nodes are randomly choosen as cluster heads and a level value 0 is assigned to them along with their ID[13].

$$CH=(ID, level\ 0)$$

During each round, each sensor node calculates its own "concentration degree" defined to be the concentration degree of node i, namely the number of sensor nodes that it can sense during the rth round.

The residual energy $E_{rd(i)}$ of a node i is calculated as $E_{rd(i)}$ = initial power – (total power consumed for transmission and reception).Sink node marks itself as level 1, and broadcast a message to neighbor nodes to construct a routing among clusters, the message is sent with "appointment of cluster heads" messages. The cluster heads one hop away from the sink will receive the message, and then marks itself as level 2 , mark the ID as its father cluster head ID. Cluster head receiving the message will mark the node ID, from which the message is sent as its father cluster ID and use the level value in the message plus 1 as its new level value if its original level value is 0. The routing construction process will end when there is no level 0 cluster head in the sensor network. In the initialization phase of network, the sink broadcasts $E_{rd(r)}$, the average energy of sensor nodes in 'cluster head election' messages. When a node i, receives the broadcast message, it will first compare its own residual energy $E_{rd(i)}$ with $E_{rd(r)}$.

if

$E_{rd(r)}<E_{rd(i)}$

{

$W_{(j,\ r)} =\alpha\ [E_{rd(i)} / E_{rd(r)}] +(1-\alpha) [D_{(j)}/ ((N-1)/K)$

$\beta=E_{rd(r)} /E_{rd(i)}$
$\alpha= 1/(1+\beta)$

}

α is adaptive factor
β is the residual energy of node j in round r
$W_{(j,r)}$= weight factor
$E_{rd(i)}$ = residual energy of node i
$E_{rd(j)}$ = average residual energy
N   = number of nodes
K   = number of clusters
    else
        Node i gives up cluster head selection and chooses to join a cluster.

The sink chooses sensor nodes with maximum election weight as cluster heads. After that, cluster heads will broadcast to neighbor nodes to notify them that it has been elected as a new cluster head. After receiving the broadcast message transmitted over a single-hop or multi-hop path, the non-cluster sensor node need to determine whether it is in the sensing range of the cluster head or not, if not they will join the CH with sensing range. Right before the beginning of a new round, source nodes should send the information of their residual energy to their cluster heads along with the sensing data for the purpose of cluster reconstruction[14][15].

4. MOBILE AGENTS IN SENSOR NETWORKS

4.1. Working of Mobile Agent

There are various model of mobile computing : client/server, peer based and agent based. In agent based modeling, a mobile agent consists of the program code and the program execution state (the current values of variables, next instruction to be executed, etc.)[16][17]. Initially a mobile agent resides on a computer called the home machine. The agent is then dispatched to execute on a remote computer called a mobile agent host (a mobile agent host is also called mobile agent platform or mobile agent server). Figure 2. describes the sequence of processes carried out during agent's lifetime. When a mobile agent is dispatched the entire code of the mobile agent(MA) and the execution state of the mobile agent is transferred to the host. The host provides a suitable execution environment for the mobile agent to execute. The mobile agent uses resources like CPU, memory, etc. of the host to perform its task[18]. After completing its task on the host, the mobile agent migrates to another computer. Since the state information is also transferred to the host, mobile agents can resume the execution of the code from where they left off in the previous host instead of having to restart execution from the beginning. This continues until the mobile agent returns to its home machine after completing execution on the last machine in its itinerary[19][20].

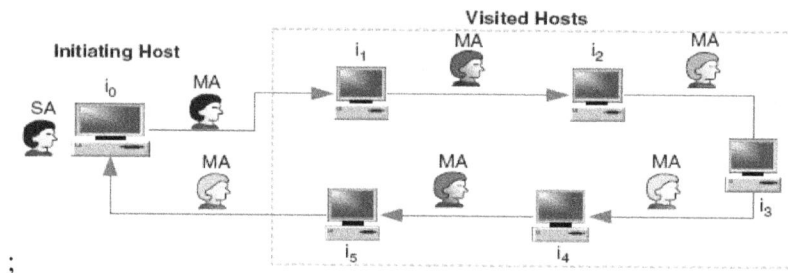

Fig 2. Agent's Lifecycle

Although, in the last several years, there has been extensive research going on in the development of mobile agent systems, there are few efforts in the study of their performance in real world applications of wireless networks. As a result, the spread of mobile agent technology in real world wireless applications cannot as yet be seen[21]. We describe the implementation of a tool to simulate the behavior of mobile agents and its utilization in industrial applications. The implementation consists of extending the existing network simulator NS2 to support mobile agent simulation.

4.2. Proposed Work

We mainly focus on the use of mobile agents in sensor networks for information retrieval and to improve the latency in the data delivery. We use the models of event driven (source initiated) and poll driven WSN model. In event driven model, the source sensor monitors for a given event (sensors are generally designed for specific applications) and reports it to base station. In a poll driven (sink initiated) model, the base station polls the source station whenever the information about a particular event is required. The architecture basically uses mobile agents ability to carry processing codes that allow the computation and communication resources at the sensor nodes to be efficiently harnessed in an undefined area. The MAs should adjust their behavior depending on quality of service needs (e.g. data delivery latency) and the network characteristics to increase network lifetime while still meeting those quality of service needs. The aim of the work would be to dedicate MAs to introduce retrieval of data from a sensor network. It also reduces the information redundancy and communication overhead at all levels so as to prolong network lifetime.

Consider a yarn break detection in a spinning machine application of a sensor network. The sensor nodes are deployed to monitor the rotation of a ring traveler which carries yarn in it. Any break in the yarn stops the rotation of the traveler. Now, the deployed sensor nodes across each spindle monitors the rotation of the traveler. If there is any break (event), it gives the report information to the cluster head. Sink sends the mobile agent to the cluster heads to gather the information.

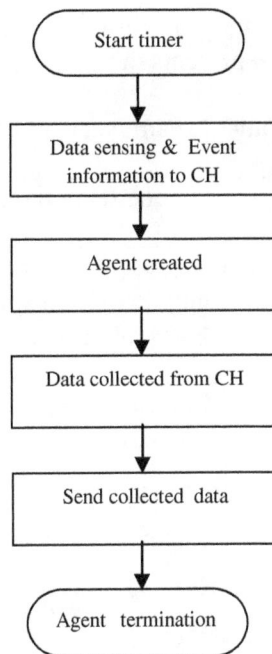

Fig 3. Flow chart representation of our model

The agent reaches the target region through multiple hops. Here it performs a three step process to retrieve and process the information: Registration- The agent checks the target node for integrity and authenticity using information being provided by the sink node. Once the authenticity check is performed, the information is collected for further processing. Association- The process of registration is followed by processing of the information gathered and removing the redundant /unwanted information. This process generally starts from second node onwards. Fusion- The process information is fused into the sink node to take appropriate action. The developed example application is structured and is shown in Figure 3. Once the timer starts, data sensing starts and the event is detected, which is then sent to cluster head, agent is created, data is collected by the agents, information is given to base station and finally the agent is terminated.

Table 1 shows the parameters of a mobile agent. The entire agent code is transmitted along with its data and execution stack. The agent's code is transmitted only if it is not available at the receiving node. The agent consists of B_{code} bytes of code, B_{data} bytes of data and B_{state} bytes of execution state and is described by the triple $B_A = (B_{code}, B_{data}, B_{state})$. The selectivity of the agent is given by σ. Thus the actual size of the results that agent is required to carry after remote processing is $(1-\sigma) B_{rep}$. The total network load for the migration of an agent A from node n1 to node n2 and then back from node n2 to node n1 is given as:

$$B_{MR} = B_{code} + B_{data} + B_{state} + (1-\sigma) B_{rep}$$

The execution time for a single agent migration from node n1 to node n2 will be

$$T_{Mig} = \delta + [(1 / \tau) + 2\mu] (B_{code} + B_{data} + B_{state})$$

The total execution time for the entire agent task, neglecting any other processing time, can be given as:

$$T_{MR} = T_{Mig} + \delta + [(1/ \tau) + 2\mu] B_{MR}$$

τ denotes the throughput of the link between nodes n1 and n2 and also μ represents the overhead. The above equation represents the mathematical model used in the simulation[22].

Table 1. Parameters of a Mobile Agent

ID	Unique ID or name given to a mobile agent.
code_size_data_size (bytes)	Represents the size of the code and data of a mobile agent respectively.
Node	Name of the current node of execution
State	Current state of the agent i.e. whether it is idle, active, mobile or terminating.
Size (bytes)	Total size of the agent migrating from one context to another.

5. SIMULATION RESULTS

We have tried to build a model for simulating the general behavior of mobile agents by extending the existing network simulator NS2. Within the clusters, the average one hop delay is calculated using the event driven MAC protocol and is plotted in Figure 4 and the number of nodes reporting an event is also represented in Figure 5. Then, the average energy consumed is calculated using the energy efficient cluster formation algorithm and is shown in Figure 6.. Using this model, the performance of the mobile agents are studied and compared with more traditional client-server approach. Figure 7 compares the time required for client-server approach(upper curve) and mobile agent based approach(lower curve).

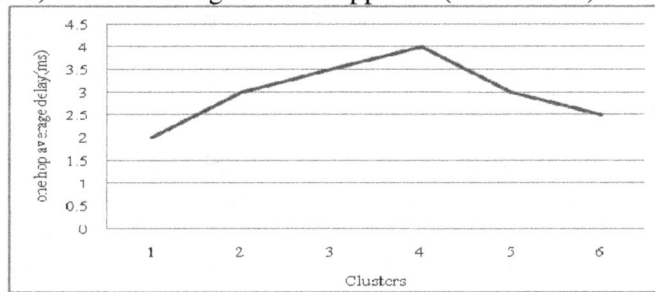

Fig 4. Number of clusters vs average one hop delay

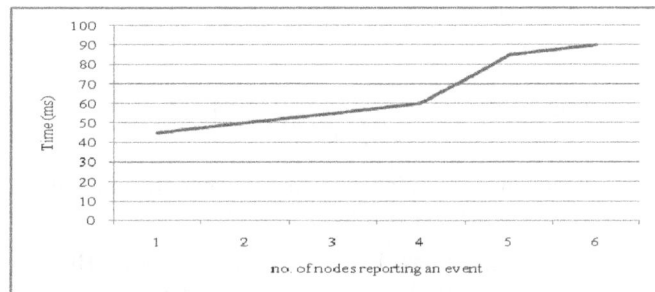

Fig 5. Number of nodes reporting an event vs time

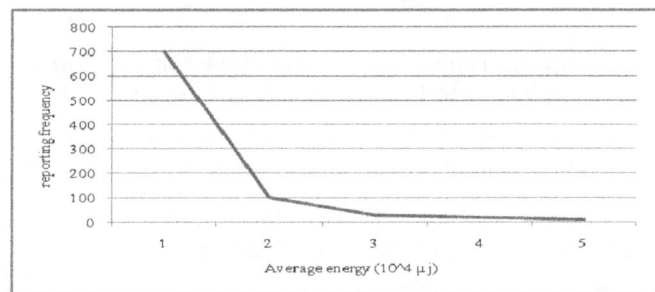

Fig 6. Energy consumption

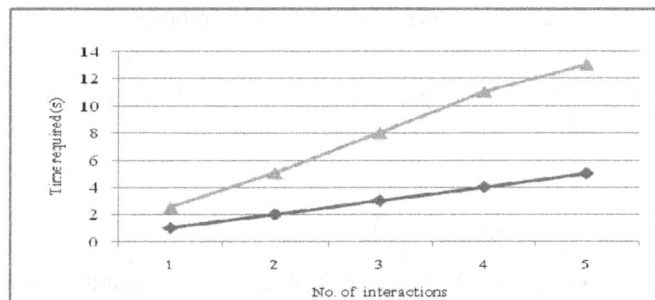

Fig 7. Effective number of Interactions

The simulation results suggest that a combination of energy efficient cluster formation, MAC layer protocol for event driven application and mobile agents to reduce the latency, may actually provide greater energy saving as well as much better latency.

6. CONCLUSION

The field of wireless sensor networks offers a rich, multi-disciplinary area of research, in which a variety of tools and concepts can be employed to address a diverse set of applications. As such, many potentials of this field have been under study both in academia and in the industry. To the best of our knowledge, we are the first to investigate the use of wireless sensor networks in spinning machine to detect the yarn break considering energy-latency problems using mobile agents. In this perspective, the results obtained from the simulation studies support the current trend of integrating the mobile agent technology into the mainstream of industrial software development rather than treating it as an exotic and specialized offshoot.

REFERENCES

[1] I.F. Akyldiz, W. Su, Y. Sankarasubramaniam and E. Cayirci, "Wireless Sensor Networks: A Survey, Computer Networks", Vol. 38, No. 4,pp. 393-422, March 2002.

[2] Ulmer C, "Organization techniques for wireless in-situ, sensor networks", http://www.craigulmer.com/research/, accessed March 20,2007.

[3]Abdelkader Outtagarts, "Mobile Agent-based Applications : a Survey", IJCSNS International Journal of Computer Science and Network Security, Vol .9 No.11, November 2009.

[4]Suriyakala C.D, Sankaranarayanan. P.E, " Smart Multiagent Architecture for Congestion Control to Access Remote Energy Meters", International Conference on Computational Intelligence and Multimedia Applications,2007.

[5] Ilker demirkol, Cem Ersoy and Fatih Alagoz, "MAC protocols for Wireless sensor networks: A Survey", IEEE Communications Magazine Topics in Ad hoc and Sensor networks, April 2006.

[6] Chung-Jung Fu, et al., "A Latency MAC Protocol for Wireless Sensor Networks", International Journal of Future Generation Communication and Networking, Vol 2, No.1, March 2009.

[7] I. Marın,et al., "LL-MAC: A low latency MAC protocol for wireless self-organised networks",Elsvier, Microprocessors and Microsystems 32 (2008) pp197–209.

[8]Lu.G, Krishnamachari.B, Raghavendra C.S., "An adaptive energy efficient and low-latency MAC for data gathering in wireless sensor networks", Proceedings of 18th International parallel Distributed Processing symposium, Pages: 224, 26-30 April 2004.

[9]Ye.W, Heiemann.J, Estrin. D, "An energy efficient MAC protocol for wireless sensor networks", IEEE Proc.Globecom 2001, pp 2944-2948.

[10]Kyle Jamieson, et al., "Sift : A MAC protocol for Event-driven Wireless sensor networks", EWSN 2006, LNCS 3868,pp 260-275.

[11]Fatma Bouabdallah, Nizar Bouabdallah, "The tradeoff between maximizing the sensor network lifetime and the fastest way to report reliably an event using reporting nodes selection", Elsevier, Computer communications 31,2008,pp 1763-1776.

[12]Yuzhong Chen and Yiping Chen, "An energy efficient clustering algorithm based on residual energy and concentration degree in wireless sensor networks", ISCSCT'09, Dec 26-28, 2009, pp306-309.

[13] Bolian Yin, Hongchi Shi, and Yi Shang, "Analysis of Energy Consumption in Clustered Wireless Sensor Networks", 2nd International symposium on Wireless Pervasive Computing, ISWPC '07.

[14] Ming Zhang and Chenglong Gong, "Energy-Efficient Dynamic Clustering Algorithm in Wireless Sensor Networks", International Symposium on Computer Science and Computational Technology,2008, pp303-306.

[15] Adeel Akhtar, et al., "Energy Aware Intra Cluster Routing for Wireless sensor networks", International Journal of Hybrid Information Technology,Vol.3, No.1, January, 2010.

[16]Francesco Aiello et al., "A Java-Based Agent Platform for Programming Wireless Sensor Networks", The Computer Journal (2010) doi: 10.1093.

[17]Abdelkader Outtagarts, "Mobile Agent based Applications -A Survey", International Journal of Computer Science and Network Security, Vol 9, No.11, November 2009.

[18] Hongjoong Sin, et al., "Agent-based Framework for Energy Efficiency in Wireless Sensor Networks", World Academy of Science, Engineering and Technology 46, 2008.

[19] Sajid Hussain et al., "Agent-based system architecture for wireless sensor networks", 20th International conference on Advanced Information Networking and Applications, April18-20, 2006.

[20] Osborne and K. Shah, "Performance Analysis of Mobile Agents in Wireless Internet Applications using simulation", A Scientific and Technical Publishing company, OACTA press.

[21]Al-Jaljouli. R, Abawajy J, "Agents Based e-Commerce and Securing Exchanged information", Pervasive computing, Springer London, 209, pp 383-404.

[22] Kunal Shah, "Performance Analysis of Mobile Analysis of Mobile Agents in wireless Internet Applications using simulation", A thesis presented to the faculty of the College of Graduate Studies, Lamar University.

4

STATISTICAL MODEL OF DOWNLINK POWER CONSUMPTION IN CELLULAR CDMA NETWORKS

Stylianos P. Savaidis[1] and Nikolaos I. Miridakis [2, 3]

[1]Department of Electronics and [2]Department of Computer Engineering, Technological

Educational Institute (TEI) of Piraeus, 250 Thivon & P.Ralli, Aigaleo, Athens–12244,

Greece

[3]Department of Informatics, University of Piraeus, 80 Karaoli & Dimitriou, 185 34

Piraeus, Greece

Abstract—*Present work proposes a theoretical statistical model of the downlink power consumption in cellular CDMA networks. The proposed model employs a simple but popular propagation model, which breaks down path losses into a distance dependent and a log-normal shadowing loss term. Based on the aforementioned path loss formalism, closed-form expressions for the first and the second moment of power consumption are obtained taking into account conditions placed by cell selection and handoff algorithms. Numerical results for various radio propagation environments and cell selection as well as handoff schemes are provided and discussed.*

Index Terms: Cellular CDMA, Downlink, Power Consumption, Soft Handoff

I. INTRODUCTION

Code division multiple access (CDMA) have been adopted by narrowband 2G and wideband 3G cellular wireless networks, due to its inherent virtue of providing a single frequency reuse pattern. Since the available spectrum is shared among all active users, the transmission power is the basic radio resource of CDMA based systems. In this context, power consumption becomes the dominant performance evaluation figure that determines network resource allocation and capacity.

Power consumption depends on the location of the mobile station (*MS*), traffic parameters and the QoS requirements of each service, experienced interference level as well as cell selection and handoff settings. Thus, the development of a power consumption model which takes into account the aforementioned parameters is a prerequisite for efficient deployment of CDMA networks. Typically, research activities on the area can be classified into those that examine the uplink [1]-[8] and the ones referring to the downlink direction [1], [3], [9]-[18]. Taking into account the asymmetric nature of data flows, the downlink is most likely to be the bottleneck point of CDMA networks. In addition, research studies of the uplink have provided analytical methodologies concluding to closed form expressions [19], which can tackle both hard and soft handoff connection modes. Typically, the downlink studies conclude to numerical simulations [1], [13], [14], [19], assumptions that simplify the examined network scenarios [3], [9], [12] or approximations that mainly resolve the complexity of calculations regarding soft handoff connection modes [9]-[11], [15]-[18]. Thus, modeling of the downlink in CDMA cellular networks is a rather important but laborious task.

Several research studies, as mentioned before, have developed an analytical methodology for the downlink performance evaluation but they resort to Monte Carlo simulations, when soft handoff is taken into account [1], [13], [14], [19]. In [3], [10] and [12] an analytical framework with closed-form expressions has been obtained but these works do not consider the soft handoff option, which requires particular attention in CDMA networks. Both hard and soft handoff connection modes are analyzed in [9] but the obtained closed-form expressions estimate the minimum downlink capacity. In [11] the complicated sums of the log-normal interferences that typically appear in soft handoff connection mode have been approximated by a log-normal distribution, which concludes to closed form expressions regarding the downlink capacity. Apart from the aforementioned approximation, the capacity evaluation in [11] simplifies the impact of soft handoff assuming that interference contributed by the soft handoff users is double, when compared with the hard handoff users. In [15] a rather efficient calculation methodology is introduced, which can estimate downlink capacity and outage probability considering both Active Set (AS) size and soft handoff option. The proposed methodology provides general analytical expressions but it demonstrates a rather high computational load, whereas the capacity calculations are possible using approximations according to the Central Limit Theory. A soft handoff scheme aiming to minimize power consumption and increase connection reliability is introduced in [16]. The proposed model in [16] approximates the sums of log-normally distributed random variables appearing in the various expressions as a single log-normal variable. Closed form expressions for the average power consumption are provided in [17] but still the numerical implementation requires a Monte Carlo simulation under soft handoff conditions and balanced power allocation for the involved Base Stations (BSs). In [18] an alternative calculation methodology is introduced in order to derive closed form expressions of the capacity at a certain outage probability. Nevertheless, the former expressions were obtained using an approximation of the energy per bit to interference ratio introducing a "macrodiversity non-orthogonality factor" and Gamma approximations of the interferences and signals in soft handoff conditions.

According to the above mentioned description, the development of a theoretical statistical model that facilitates performance evaluation of the downlink in cellular CDMA networks becomes quite laborious, especially when soft handoff is considered. Approximating assumptions or numerical simulations are typically employed in order to overcome the complexity of analysis. The presence of sums of log-normally distributed random variables in the various expressions is the major obstacle regarding the derivation of closed-form analytical expressions. The present work proposes an alternative approach in order to overcome this kind of complexity and conclude to closed form expressions. In particular, a Taylor series expansion of the aforementioned complicated expressions is employed, which next makes possible a straightforward calculation of power consumption moments. In fact present work demonstrates the calculation procedure for the first two moments of power consumption, although in principle affords calculation of higher order moments. The proposed calculation scheme can integrate several realistic conditions including a best BS selection condition, the impact of a soft handoff threshold as well as AS size. Overall, the present work provides a theoretical statistical model, which attempts to balance efficiently between the assumptions that oversimplify the examined network scenarios, the inaccuracies of the potential approximations and the physical insight that a closed form expression may provide.

Section II, describes the radio propagation model and the downlink power consumption formulas for hard handoff (*HHO*), 2-way and 3-way soft handoff (*SHO*) connection modes. Section III describes the conditions placed by cell selection and handoff schemes. In Section IV, the calculation details for the first and second moments of the downlink power consumption are discussed in details. Section V includes numerical results and verification tests regarding the proposed calculation scheme. Finally, section VI summarizes the main conclusions and discusses potential extensions of current work.

II. DOWNLINK POWER CONSUMPTION

The adopted radio propagation model assumes that fast fading can be compensated by special reception techniques, e.g. rake receiver, thus it can be considered as a pure large scale path loss model. In particular, path losses are solely determined by a path loss factor, which determines the distance based losses, and a shadowing loss component, which demonstrates a log-normal behavior. Thus, the power received from a transmitting BS can be determined by the following expression:

$$P(r, \zeta) = r^{-\alpha} 10^{\zeta/10} P_T \tag{1}$$

where r denotes the distance between *MS* and *BS* and P_T the *BS*'s total transmitted power; α is the path loss factor and ζ denotes the shadowing losses as a zero-mean Gaussian distributed random variable with standard deviation σ. The shadowing loss random variable for a certain *BS*, i.e. BS_i, can be further analyzed into two components, namely $\zeta_i = a\xi + b\xi_i$ [13], [14]. The $a\xi$ component denotes a part of shadowing that is common for all *BSs* and it represents the environment near and around the *MS*, whereas $b\xi_i$ denotes shadowing effects that depend on the environment near and around *BS*. The constants a and b, fulfill the relationship $a^2+b^2=1$, whereas ξ_i are considered as independent zero-mean Gaussian distributed random variables with standard deviation σ [12]-[14].

The network scenario under investigation considers center feed cells of hexagonal shape and equal size. The interference and downlink power consumption analysis assumes an *MS*, which camps in cell 1 with two tiers of neighboring cells around it, as Fig. 1 depicts. Intra-cell interference calculations require only the knowledge of the distance r_1 between the serving BS_1 and the *MS*. However, for inter-cell interference calculations, both distance r_1 and angular position θ_1, as Fig. 2 shows, should be considered. The distance r_1 between MS and BS$_1$ varies from zero to $\sqrt{3}R\cos(\theta_1)/2$, whereas the angular coordinate θ_1 varies from 0° to 360°. Due to the hexagonal symmetry, throughout the remaining analysis only angular positions $\theta_1=0° \sim 30°$, will be examined.

Power control function should under ideal conditions regulate downlink power consumption in order to lock energy per bit to interference value to the target value $[E_b/I_o]_t$ required by each service. Thus, by calculating interference level and assuming a perfect power control scheme, downlink power consumption for *HHO*, 2-way and 3-way *SHO* connection modes can be estimated as follows:

A) Hard Handoff Scenario

When a *HHO* connection mode is assumed, all downlink transmissions to other *MS*s within the cell as well as in neighbor cells are considered interference. In principle, the proposed model can tackle network scenarios with unequal traffic loads per cell and thus different transmit power level $P_{T,i} = \delta_i P_T$ per base station. However, in order to simplify model's analysis, we assume equal total downlink transmission levels P_T in each cell (i.e. $\delta_i = 1$). In this respect, power consumption for a single connection in cell 1 can be calculated as follows [13]-[14]:

$$\left[\frac{E_b}{I_o}\right]_t = \frac{W}{vR} \frac{r_1^{-\alpha} 10^{\zeta_1/10} P_{s1}}{(1\text{-}u) r_1^{-\alpha} 10^{\zeta_1/10}(P_T - P_{s1}) + \sum_{i=2}^{19} r_i^{-\alpha} 10^{\zeta_i/10} P_T} \Rightarrow$$

$$\Rightarrow P_{s1} = C_t X(\underline{\xi}) P_T = C_t \left[\sum_{i=0,i\neq1}^{19} X_i\right] P_T = \beta_1 P_T \qquad (2)$$

where $C_t = vR[E_b/I_o]_t/W$, $X_0 = 1-u$ denotes intra-cell interference, $X_i = C_{1,i} 10^{b(\xi_i - \xi_1)/10}$ denotes inter-cell interference and $C_{1,i} = (r_1/r_i)^\alpha$. Vector $\underline{\xi} = (\xi_1, ..., \xi_{19})$ denotes the uncorrelated shadowing random variables of BS_is, v is the activity factor which applies to the service under examination, R is the service data rate, W is the chip rate, u denotes the orthogonality between the various transmissions and β_1 is the fraction of P_T allocated for a single link. For the sake of simplicity, in equation (2) and throughout equations (3) and (4), we assume that $(P_T - P_{s1}) \cong P_T$ as far as it concerns intracell interference calculations.

B) 2-way Soft Handoff Scenario

If we assume a maximal ratio combination capability (MRC) and a balanced power allocation scheme ($P_{s1} = P_{sk} = P_{s,1k}$) among BSs in cell 1 and cell k, which participate in the 2-way SHO connection, then power consumption for a single connection in cell 1 is calculated as follows [9], [13]-[14]:

$$\left[\frac{E_b}{I_o}\right]_t = \left[\frac{E_b}{I_o}\right]_1 + \left[\frac{E_b}{I_o}\right]_k = \frac{W}{vR}\left[\frac{r_1^{-\alpha} 10^{\zeta_1/10} P_{s1}}{(1-u) r_1^{-\alpha} 10^{\zeta_1/10} P_T + \sum_{i=2}^{19} r_i^{-\alpha} 10^{\zeta_i/10} P_T} + \right.$$

$$\left. + \frac{r_k^{-\alpha} 10^{\zeta_k/10} P_{sk}}{(1-u) r_k^{-\alpha} 10^{\zeta_k/10} P_T + \sum_{i=1,i\neq k}^{19} r_i^{-\alpha} 10^{\zeta_i/10} P_T}\right] \Rightarrow P_{s,1k} = C_t\left[\frac{1}{X(\underline{\xi})} + \frac{1}{Y(\underline{\xi})}\right] P_T =$$

$$= C_t\left[\left(\sum_{i=0,i\neq1}^{19} X_i\right)^{-1} + \left(\sum_{i=0,i\neq k}^{19} Y_i\right)^{-1}\right] P_T = \beta_{1k} P_T \qquad (3)$$

where similar to eq. (2), $Y_0 = 1-u$, $Y_i = C_{k,i} 10^{b(\xi_i - \xi_k)/10}$ and $C_{k,i} = (r_k/r_i)^\alpha$; β_{1k} is the fraction of P_T allocated by each BS, which participates to the 2-way SHO connection.

C) 3-way Soft Handoff Scenario

If we assume MRC reception conditions and balanced power allocation scheme ($P_{s1} = P_{sk} = P_{sl} = P_{s,1kl}$) between *BS*s in cell 1, cell *k* and cell *l*, which participate in the 3-way SHO connection, then power consumption for a single connection in cell 1 is calculated as in the previous cases [9], [13]-[14]:

$$\left[\frac{E_b}{I_o}\right]_t = \left[\frac{E_b}{I_o}\right]_1 + \left[\frac{E_b}{I_o}\right]_k + \left[\frac{E_b}{I_o}\right]_l = \frac{W}{vR}\left[\frac{r_1^{-\alpha}10^{\zeta_1/10}P_{s1}}{(1-u)r_1^{-\alpha}10^{\zeta_1/10}P_T + \sum_{i=2}^{19}r_i^{-\alpha}10^{\zeta_i/10}P_T} + \right.$$

$$\left. + \frac{r_k^{-\alpha}10^{\zeta_k/10}P_{sk}}{(1-u)r_k^{-\alpha}10^{\zeta_k/10}P_T + \sum_{i=1,i\neq k}^{19}r_i^{-\alpha}10^{\zeta_i/10}P_T} + \frac{r_l^{-\alpha}10^{\zeta_l/10}P_{sl}}{(1-u)r_l^{-\alpha}10^{\zeta_l/10}P_T + \sum_{i=1,i\neq l}^{19}r_i^{-\alpha}10^{\zeta_i/10}P_T}\right] \Rightarrow$$

$$\Rightarrow P_{s,1kl} = C_t\left[\frac{1}{X(\underline{\xi})} + \frac{1}{Y(\underline{\xi})} + \frac{1}{Z(\underline{\xi})}\right]P_T = C_t\left[\left(\sum_{i=0,i\neq1}^{19}X_i\right)^{-1} + \left(\sum_{i=0,i\neq k}^{19}Y_i\right)^{-1} + \left(\sum_{i=1,i\neq l}^{19}Z_i\right)^{-1}\right]P_T = \beta_{1kl}P_T \quad (4)$$

where similar to eq. (2)-(3), $Z_0 = 1-u$, $Z_i = C_{l,i}10^{b(\xi_i-\xi_l)/10}$ and $C_{l,i} = (r_l/r_i)^{\alpha}$. β_{1kl} is the fraction of P_T allocated by each BS, which participates to the 3-way *SHO* connection.

At this point it should be mentioned that in principle, the proposed model can tackle both balanced and unbalanced power allocation schemes by defining different weights on P_{s1}, P_{sk} and P_{sl}. Nevertheless, for simplicity reasons, in our analysis we assume equal weights on P_{s1} P_{sk} and P_{sl}, yet without loss of generality.

III. CELL SELECTION AND HANDOFF SCHEMES

Cell selection and handover schemes influence the network performance [20], [21] and thus current section will examine the conditions that are imposed in our calculations by the aforementioned schemes. If cell 1 is the camping cell and assuming a best BS selection condition, then the transmission of cell-1 will be the best among the candidate cells *i* (=2, 3,..., 19), i.e. $\xi_i \leq \xi_1 - R_{1,i}$ ($R_{1,i} = 10\log(C_{1,i})/b$). The former condition describes an ideal cell selection scenario and a perfect power control scheme. The addition of a hysteresis threshold *cst*=10log(*CST*)/*b* can account for possible cell selection and power control imperfections e.g. $\xi_i \leq \xi_1 - R_{1,i} + cst$.

Apart from the above described conditions, the handoff algorithm is placing additional ones. The handoff scheme considered here is one that accepts a maximum number of simultaneous physical connections equal to the *AS* size. In addition, the algorithm places a *SHO* threshold in order to accept a *BS* to join the *AS*. If *SHO* is not an option, i.e. *AS*=1, then the handoff condition is identical to the cell selection one. However, if *AS*>1, then the *HHO* scenario implies that the signal strength of all monitored *BS*s should not exceed the *SHO* threshold. The latter statement is expressed as $\xi_i \leq \xi_1 - R_{1,i} - sht$, $(sht = 10\log(SHT)/b)$. Concluding with the *HHO* mode, the following conditions apply:

$$\xi_i \leq \xi_1 - R_{1,i} - sht, \; AS > 1 \tag{5}$$

$$\xi_i \leq \xi_1 - R_{1,i} + cst, \; AS = 1 \tag{6}$$

If 2-way *SHO* conditions apply, then two simultaneous connections with BS_1 and BS_k occur. If $AS>1$, then BS_k's signal is the strongest signal among the monitored ones and exceeds *SHO* threshold. After some straightforward calculations, the former statements can be described as follows:

$$\xi_1 - R_{1,k} - sht \leq \xi_k \leq \xi_1 - R_{1,k} + cst, \; k \in AS > 1 \tag{7}$$

$$\xi_i \leq \xi_k - R_{k,i}, \; i \notin AS (= 2) \tag{8}$$

$$\xi_i \leq \xi_1 - R_{1,i} - sht, \; i \notin AS (= 3) \tag{9}$$

Finally, when $AS=3$ a 3-way *SHO* scenario applies and a single logical network link include physical links with three *BS*s, e.g. BS_1, BS_k and BS_l. BS_k's and BS_l's signal are the strongest signals among the monitored ones and both exceed the *SHO* threshold. Assuming that BS_l's signal is the weakest among the *AS* participants, then all other monitored signals should be weaker than BS_l's signal. After some straightforward calculations the former statements can be expressed as follows:

$$\xi_1 - R_{1,k(l)} - sht \leq \xi_{k(l)} \leq \xi_1 - R_{1,k(l)} + cst, \; k(l) \in AS \tag{10}$$

$$\xi_l \leq \xi_k - R_{k,l}, \; k \; and \; l \in AS \tag{11}$$

$$\xi_i \leq \xi_l - R_{l,i}, \; i \notin AS \tag{12}$$

Concluding, it is worthwhile to mention that no restrictions are placed in non monitored cells, which *de facto* do not participate to handoff process. In order to simplify the analysis throughout the remaining analysis all cells in both tiers will be considered as monitored.

IV. DOWNLINK POWER CONSUMPTION STATISTICS

Three handoff schemes are considered in this section, i.e. $AS=1$, 2 and 3. In all following calculations, the random shadowing loss values ξ_i are restricted by the cell selection and handoff conditions discussed in the previous section. If $AS=m$ and Ω^m is the subset of random ξ_i values, which allow *MS* to camp in cell 1, then Ω^m can be expressed as $\Omega^m = \Omega_1^m \cup \sum_k \Omega_{1k}^m \cup \sum_{k,l} \Omega_{1kl}^m$. Subsets Ω_1^m, Ω_{1k}^m and Ω_{1kl}^m include all ξ_i values, which conform to *HHO*, 2-way *SHO* and 3-way *SHO* conditions, respectively. The conditions for each subset are established with eqs. (5)-(6), (7)-(9) and (10)-(12) of section III. Apparently $\Omega_{1k}^1 = \varnothing$ and $\Omega_{1kl}^1 = \Omega_{1kl}^2 = \varnothing$.

The above discussed subsets correspond to all possible connection modes that may occur in the cell under investigation i.e. cell 1. If the downlink transmitted power for a single user in cell 1 is $P_s = \beta P_T$, then the actual point of interest in our calculations is the fraction β of the total transmitted power. The first and the second moment of β can be obtained as

$$E[\beta \mid \underline{\xi} \in \Omega^m] = \frac{P(\Omega_1^m)E[\beta_1 \mid \underline{\xi} \in \Omega_1^m] + \sum_k P(\Omega_{1k}^m)E[\beta_{1k} \mid \underline{\xi} \in \Omega_{1k}^m] + \sum_{k,l} P(\Omega_{1kl}^m)E[\beta_{1kl} \mid \underline{\xi} \in \Omega_{1kl}^m]}{P(\Omega^m)} \quad (13)$$

$$E[\beta^2 \mid \underline{\xi} \in \Omega^m] = \frac{P(\Omega_1^m)E[\beta_1^2 \mid \underline{\xi} \in \Omega_1^m] + \sum_k P(\Omega_{1k}^m)E[\beta_{1k}^2 \mid \underline{\xi} \in \Omega_{1k}^m] + \sum_{k,l} P(\Omega_{1kl}^m)E[\beta_{1kl}^2 \mid \underline{\xi} \in \Omega_{1kl}^m]}{P(\Omega^m)} \quad (14)$$

where $P(\Omega^m) = P(\Omega_1^m) + \sum_k P(\Omega_{1k}^m) + \sum_{k,l} P(\Omega_{1kl}^m)$.

Since the shadowing random variables ξ_i are independent their joint pdf is

$$f_{\underline{\xi}}(\underline{\xi}) = \prod_{n=1}^{19} f_{\xi_i}(\xi_i) = \prod_{n=1}^{19} \frac{e^{(\xi_i^2/2\sigma^2)}}{\sqrt{2\pi}\sigma} \quad (15)$$

and thus $P(\Omega_1^m)$ can be calculated as follows:

$$P(\Omega_1^m) = \int_{-\infty}^{+\infty} f_{\xi_1}(\xi_1) \left\{ \prod_{n=2}^{19} \int_{-\infty}^{a_n(\xi_1)} f_{\xi_n}(\xi_n) \right\} d\xi_1 = \int_{-\infty}^{+\infty} f_{\xi_1}(\xi_1) \left\{ \prod_{n=2}^{19} A(a_n(\xi_1), 0) \right\} d\xi_1 \quad (16)$$

where we define function $A(x,y)$ as

$$A(x, y) = \exp\left[y^2 \sigma^2 b^2 \ln(10)^2/200 \right] \left\{ 0.5 + 0.5 erf \left[\frac{x}{\sigma\sqrt{2}} - y \frac{\sigma b \ln(10)}{10\sqrt{2}} \right] \right\} \quad (17)$$

and $a_n(\xi_1)$ is the upper limit of inequality (5) or (6), when $m > 1$ or $m=1$, respectively. In a similar manner $P(\Omega_{1k}^m)$ is obtained by the following expressions:

$$P(\Omega_{1k}^m) = \int_{-\infty}^{+\infty} f_{\xi_1}(\xi_1) \left\{ \int_{b_k(\xi_1)}^{a_k(\xi_1)} f_{\xi_k}(\xi_k) \left[\prod_{n=2, n\neq k}^{19} A(a_n(\xi_k), 0) \right] d\xi_K \right\} d\xi_1 \quad (18)$$

where $a_k(\xi_1)$ and $b_k(\xi_1)$ is the upper and the lower limit of inequality (7), respectively. If $m=2$ then $a_n(\xi_k)$ is the upper limit of inequality (8), otherwise $a_n(\xi_k)(=a_n(\xi_1))$ is the upper limit of equation (9). In addition, if $m=2$ the integration over ξ_k can be only evaluated numerically, whereas for $m=3$ the integration over ξ_k is evaluated analytically as $[A(b_k(\xi_1),0)-A(a_k(\xi_1),0)]$. With a similar manipulation $P(\Omega_{1kl}^m)$ is obtained by the following expression:

$$P(\Omega_{1kl}^m) = \int_{-\infty}^{+\infty} f_{\xi_1}(\xi_1) \left\{ \int_{b_k(\xi_1)}^{a_k(\xi_1)} f_{\xi_k}(\xi_k) \left(\int_{b_l(\xi_k)}^{a_l(\xi_k)} f_{\xi_l}(\xi_l) \left[\prod_{n=2, n\neq k,l}^{19} A_n(a_n(\xi_l),0) \right] d\xi_l \right) d\xi_k \right\} d\xi_1 \quad (19)$$

where $a_k(\xi_l)$ and $b_k(\xi_1)$ is the upper and the lower limit of eq. (10), $a_l(\xi_k)$ and $b_l(\xi_1)$ is the upper and the lower limit of eqs. (11) and (10), respectively, whereas $a_n(\xi_l)$ is the upper limit of eq. (12).

A) HHO Calculations

According to equation (2) the first and the second moments of β_1 can be obtained as follows:

$$E[\beta_1] = C_t \sum_{\substack{i=0 \\ i\neq1}}^{19} E[X_i], \ E[\beta_1^2] = C_t^2 \left\{ \sum_{\substack{i=0 \\ i\neq1}}^{19} E[X_i^2] + \sum_{\substack{j=0 \\ j\neq1}}^{19} \sum_{\substack{i=0 \\ i\neq1, i\neq j}}^{19} E[X_j X_i] \right\} \quad (20)$$

where for $i\neq0$

$$E[X_i] = \frac{C_{1,i}}{P(\Omega_1^m)} \int_{-\infty}^{+\infty} 10^{-b\xi_1/10} f_{\xi_1}(\xi_1) A(a_i(\xi_1),1) \Pi_i(\xi_1) d\xi_1 \quad (21)$$

$$E\left[X_i^2\right] = \frac{C_{1,i}^2}{P(\Omega_1^m)} \int_{-\infty}^{+\infty} 10^{-b\xi_1/5} f_{\xi_1}(\xi_1) A(a_i(\xi_1),2) \Pi_i(\xi_1) d\xi_1 \quad (22)$$

$$E\left[X_i X_j\right] = \frac{C_{1,i}C_{1,j}}{P(\Omega_1^m)} \int_{-\infty}^{+\infty} 10^{-b\xi_1/5} f_{\xi_1}(\xi_1) A(a_i(\xi_1),1) A(a_j(\xi_1),1) \Pi_{i,j}(\xi_1) d\xi_1 \quad (23)$$

and $\Pi_i(\xi_1) = \prod_{n=2,n\neq i}^{19} A(a_n(\xi_1),0)$, $\Pi_{i,j}(\xi_1) = \prod_{n=2,n\neq i,j}^{19} A(a_n(\xi_1),0)$. In addition, $E[X_0]=(1-u)/P(\Omega_1^m)$, $E[X_0^2]=(1-u)^2/P(\Omega_1^m)$, $E[X_0X_j]=(1-u)E[X_j]$ with $E[X_j]$ given by (21) if we replace j with i. The integration limits of the above expressions are the same with the ones appearing in eq. (16).

B) 2-way SHO Calculations

According to eq. (3) the first and the second moment of $\beta_{1,k}$, , can not be evaluated by employing the straightforward semi-analytical approach of subsection IV.A. In order to overcome this constraint, β_{1k} is approximated by a Taylor expansion in the neighborhood of $E[X(\underline{\xi})]$ and $E[Y(\underline{\xi})]$. Next, by omitting Taylor series terms higher than the second order we conclude to (see Appendix I):

$$E[\beta_{1k}\mid\underline{\xi}\in\Omega_{1k}^m] = C_t \left\{ \frac{\overline{XY}}{\overline{X}+\overline{Y}} - \frac{\left[\overline{X^2}-\left(\overline{X}\right)^2\right]\left(\overline{Y}\right)^2 + \left[\overline{Y^2}-\left(\overline{Y}\right)^2\right]\left(\overline{X}\right)^2 - 2\left[\overline{XY}-\overline{X}\,\overline{Y}\right]\overline{X}\,\overline{Y}}{\left[\overline{X}+\overline{Y}\right]^3} \right\} \quad (24)$$

$$E[\beta_{1k}^2\mid\underline{\xi}\in\Omega_{1k}^m] = C_t^2 \left\{ \frac{\left(\overline{Y}\right)^4\overline{X^2}+\left(\overline{X}\right)^4\overline{Y^2}+2\left(\overline{XY}\right)^2\overline{XY}}{\left[\overline{X}+\overline{Y}\right]^4} + 2\frac{\overline{XY}}{\overline{X}+\overline{Y}}\left[\frac{E[\beta_{1k}\mid\underline{\xi}\in\Omega_{1k}^m]}{C_t} - \frac{\overline{XY}}{\overline{X}+\overline{Y}} \right] \right\} (25)$$

where $\overline{X}$, $\overline{Y}$, $\overline{X^2}$, $\overline{Y^2}$ and $\overline{XY}$ correspond to $E[X\mid\underline{\xi}\in\Omega_{1k}^m]$, $E[Y\mid\underline{\xi}\in\Omega_{1k}^m]$, $E[X^2\mid\underline{\xi}\in\Omega_{1k}^m]$, $E[Y^2\mid\underline{\xi}\in\Omega_{1k}^m]$ and $E[XY\mid\underline{\xi}\in\Omega_{1k}^m]$, respectively. Following a calculation scheme as in section IV.A, the above mentioned E[.] terms can be expressed as a summation of all possible combinations of $E[X_i]$, $E[X_i^2]$, $E[X_iX_j]$, $E[Y_i]$, $E[Y_i^2]$, $E[Y_iY_j]$] and $E[X_iY_j]$. Also, each E[.] term can be expressed in an integral closed form expression, where the various integration limits are identical to the ones appearing in eq. (18). The $E[X_i]$ expression is obtained as follows:

$$E[X_i] = \frac{C_{1,i}}{P(\Omega_{1k}^m)} \int_{-\infty}^{+\infty} 10^{-\frac{b\xi_1}{10}} f_{\xi_1}(\xi_1)d\xi_1 \times \begin{cases} \int_{b_k(\xi_1)}^{a_k(\xi_1)} f_{\xi_k}(\xi_k)A(a_i(\xi_k),1)\Pi_{i,k}(\xi_k)d\xi_k, i \neq k \\ \int_{b_k(\xi_1)}^{a_k(\xi_1)} 10^{\frac{b\xi_k}{10}} f_{\xi_k}(\xi_k)\Pi_k(\xi_k)d\xi_k, i = k \end{cases} \quad (26)$$

If $m=3$ and $i \neq k$ ($i=k$) the kth integral in eq. (26) can be evaluated as $[A(b_k(\xi_1),0)-A(a_k(\xi_1),0)]$ $([A(b_k(\xi_1),1)-A(a_k(\xi_1),1)])$. The $E[Y_i]$ calculations are similar to eq. (26) with one difference, namely, the term $10^{-b\xi_1/10}$ is transferred to the kth integral as $10^{-b\xi_k/10}$

$$E[Y_i] = \frac{C_{k,i}}{P(\Omega_{1k}^m)} \times \begin{cases} \int_{-\infty}^{+\infty} f_{\xi_1}(\xi_1)d\xi_1 \int_{b_k(\xi_1)}^{a_k(\xi_1)} 10^{-\frac{b\xi_k}{10}} f_{\xi_k}(\xi_k)A(a_i(\xi_k),1)\Pi_{i,k}(\xi_k)d\xi_k, i \neq 1 \\ \int_{-\infty}^{+\infty} 10^{\frac{b\xi_1}{10}} f_{\xi_1}(\xi_1)d\xi_1 \int_{b_k(\xi_1)}^{a_k(\xi_1)} 10^{-\frac{b\xi_k}{10}} f_{\xi_k}(\xi_k)\Pi_k(\xi_k)d\xi_k, i = 1 \end{cases} \quad (27)$$

If $m=3$ the kth integral in eq. (27) can be evaluated as $[A(b_k(\xi_1),-1)-A(a_k(\xi_1),-1)]$. Similar to the *HHO* case $E[X_0]=E[Y_0]=(1-u)/P(\Omega_{1k}^m)$.

The $E[X_i^2]$ and $E[Y_i^2]$ expressions can be obtained from eqs. (26) and (27), respectively, if we substitute $10^{\pm b\xi_1/10}$ and $10^{\pm b\xi_k/10}$ with $10^{\pm b\xi_1/5}$ and $10^{\pm b\xi_k/5}$, respectively, $A(a_i(\xi_k),1)$ with $A(a_i(\xi_k),2)$ and $[A(b_k(\xi_1),\pm1)-A(a_k(\xi_1),\pm1)]$ with $[A(b_k(\xi_1),\pm2)-A(a_k(\xi_1),\pm2)]$. Similar to the *HHO* case $E[X_0^2]=E[Y_0^2]=(1-u)^2/P(\Omega_{1k}^m)$.

The terms $E[X_iX_j]$ are described by the following integral expression:

$$E[X_iX_j] = \frac{C_{1,i}C_{1,j}}{P(\Omega_{1k}^m)} \int_{-\infty}^{+\infty} 10^{-\frac{b\xi_1}{5}} f_{\xi_1}(\xi_1)d\xi_1 \times \begin{cases} \int_{b_k(\xi_1)}^{a_k(\xi_1)} f_{\xi_k}(\xi_k)A(a_i(\xi_k),1)A(a_j(\xi_k),1)\Pi_{i,j,k}(\xi_k)d\xi_k, i,j \neq k \\ \int_{b_k(\xi_1)}^{a_k(\xi_1)} 10^{\frac{b\xi_k}{10}} f_{\xi_k}(\xi_k)A(a_{i(j)}(\xi_k),1)\Pi_{i,j,k}(\xi_k)d\xi_k, i(j) = k \end{cases} \quad (28)$$

where $\Pi_{i,j,k}(\xi_k) = \prod_{n=2,n \neq i,j,k}^{19} A(a_n(\xi_k),0)$. If $j=3$ and $i,j \neq k$ (*i or j=k*) the kth integral in eq. (28) can be evaluated as $[A(b_k(\xi_1),0)-A(a_k(\xi_1),0)]$ $([A(b_k(\xi_1),1)-A(a_k(\xi_1),1)])$.

The expression for the $E[Y_iY_j]$ term is given by the following equation:

$$E\left[Y_iY_j\right]=\frac{C_{k,i}C_{k,j}}{P(\Omega_{1k}^m)}\times\begin{cases}\int\limits_{-\infty}^{+\infty}f_{\xi_1}(\xi_1)d\xi_1\int\limits_{b_k(\xi_1)}^{a_k(\xi_1)}10^{-\frac{b\xi_k}{5}}f_{\xi_k}(\xi_k)A(a_i(\xi_k),1)A(a_j(\xi_k),1)\Pi_{i,j,k}(\xi_k)d\xi_k,i,j\neq1\\[2mm]\int\limits_{-\infty}^{+\infty}10^{\frac{b\xi_1}{10}}f_{\xi_1}(\xi_1)d\xi_1\int\limits_{b_k(\xi_1)}^{a_k(\xi_1)}10^{-\frac{b\xi_k}{5}}f_{\xi_k}(\xi_k)A(a_{i(j)}(\xi_k),1)\Pi_{i,j,k}(\xi_k)d\xi_k,i(j)=1\end{cases}\tag{29}$$

If $m=3$ the kth integral in eq. (29) can be evaluated as $[A(b_k(\xi_1),-2)-A(a_k(\xi_1),-2)]$.

Concluding the 2-way *SHO* subsection the $E[X_iY_j]$ term is expressed below:

$$E\left[X_iY_j\right]=\frac{C_{1,i}C_{k,j}}{P(\Omega_{1k}^m)}\times\begin{cases}\int\limits_{-\infty}^{+\infty}10^{-\frac{b\xi_1}{10}}f_{\xi_1}(\xi_1)d\xi_1\int\limits_{b_k(\xi_1)}^{a_k(\xi_1)}10^{-\frac{b\xi_k}{10}}f_{\xi_k}(\xi_k)A(a_i(\xi_k),1)A(a_j(\xi_k),1)\Pi_{i,j,k}(\xi_k)d\xi_k,\begin{Bmatrix}i\neq k,\\j\neq1\end{Bmatrix}\\[2mm]\int\limits_{-\infty}^{+\infty}10^{-\frac{b\xi_1}{10}}f_{\xi_1}(\xi_1)d\xi_1\int\limits_{b_k(\xi_1)}^{a_k(\xi_1)}f_{\xi_k}(\xi_k)A(a_j(\xi_k),1)\Pi_{i,j,k}(\xi_k)d\xi_k,\{i=k,j\neq1\}\\[2mm]\int\limits_{-\infty}^{+\infty}f_{\xi_1}(\xi_1)d\xi_1\int\limits_{b_k(\xi_1)}^{a_k(\xi_1)}10^{-\frac{b\xi_k}{10}}f_{\xi_k}(\xi_k)A(a_i(\xi_k),1)\Pi_{i,j,k}(\xi_k)d\xi_k,\{i\neq k,j=1\}\\[2mm]\int\limits_{-\infty}^{+\infty}f_{\xi_1}(\xi_1)d\xi_1\int\limits_{b_k(\xi_1)}^{a_k(\xi_1)}f_{\xi_k}(\xi_k)\Pi_{i,j,k}(\xi_k)d\xi_k,\{i=k,j=1\}\end{cases}\tag{30}$$

If $i=j\neq1$ and k, $A(a_i(\xi_k),1)A(a_j(\xi_k),1)$ product should be replaced by $A(a_i(\xi_k),2)$. Finally, if $m=3$ and $i\neq k$ $(i=k)$ the kth integral in eq. (30) is evaluated as $[A(b_k(\xi_1),-1)-A(a_k(\xi_1),-1)]$ $([A(b_k(\xi_1),0)-A(a_k(\xi_1),0)])$.

C) 3-way SHO Calculations

As it was discussed in the 2-way *SHO* case the first and the second moment of β_{1kl}, can be approximated through a Taylor expansion of eq. (4). If we omit Taylor series terms higher than the second order the following expressions can be derived for the first and the second moment of β_{1kl} (see Appendix II):

$$E[\beta_{1kl}\mid\underline{\xi}\in\Omega_{1kl}^m]=C_t\left\{\frac{\overline{XYZ}}{\overline{XY}+\overline{XZ}+\overline{YZ}}-\frac{2\left[\overline{X^2}-(\overline{X})^2\right]\left[(\overline{Y}+\overline{Z})(\overline{YZ})^2\right]}{\left[\overline{XY}+\overline{XZ}+\overline{YZ}\right]^3}-\right.$$

$$-\frac{2\left[\overline{Y^2}-(\overline{Y})^2\right]\left[(\overline{X}+\overline{Z})(\overline{XZ})^2\right]}{\left[\overline{XY}+\overline{XZ}+\overline{YZ}\right]^3}-\frac{2\left[\overline{Z^2}-(\overline{Z})^2\right]\left[(\overline{X}+\overline{Y})(\overline{XY})^2\right]}{\left[\overline{XY}+\overline{XZ}+\overline{YZ}\right]^3}+\frac{2\left[\overline{XY}-\overline{X}\,\overline{Y}\right]\overline{XY}(\overline{Z})^3}{\left[\overline{XY}+\overline{XZ}+\overline{YZ}\right]^3}+$$

$$\left.+\frac{2\left[\overline{XZ}-\overline{X}\,\overline{Z}\right]\overline{XZ}(\overline{Y})^3}{\left[\overline{XY}+\overline{XZ}+\overline{YZ}\right]^3}+\frac{2\left[\overline{YZ}-\overline{Y}\,\overline{Z}\right]\overline{YZ}(\overline{X})^3}{\left[\overline{XY}+\overline{XZ}+\overline{YZ}\right]^3}\right\}\tag{31}$$

$$E[\beta_{1kl}^2 \mid \underline{\xi} \in \Omega_{1kl}^m] = C_t^2 \left\{ \left(\frac{\overline{XYZ}}{\overline{XY} + \overline{XZ} + \overline{YZ}} \right)^2 + \frac{\left[\overline{X^2} - (\overline{X})^2\right](\overline{YZ})^4}{\left[\overline{XY} + \overline{XZ} + \overline{YZ}\right]^4} + \frac{\left[\overline{Y^2} - (\overline{Y})^2\right](\overline{XZ})^4}{\left[\overline{XY} + \overline{XZ} + \overline{YZ}\right]^4} + \right. $$

$$+ \frac{\left[\overline{Z^2} - (\overline{Z})^2\right](\overline{XY})^4}{\left[\overline{XY} + \overline{XZ} + \overline{YZ}\right]^4} + \frac{2\left[\overline{XY} - \overline{XY}\right](\overline{XY})^2(\overline{Z})^4}{\left[\overline{XY} + \overline{XZ} + \overline{YZ}\right]^4} + \frac{2\left[\overline{XZ} - \overline{XZ}\right](\overline{XZ})^2(\overline{Y})^4}{\left[\overline{XY} + \overline{XZ} + \overline{YZ}\right]^4} + $$

$$+ \frac{2\left[\overline{YZ} - \overline{YZ}\right](\overline{YZ})^2(\overline{X})^4}{\left[\overline{XY} + \overline{XZ} + \overline{YZ}\right]^4} + \frac{2\overline{XYZ}}{\overline{XY} + \overline{XZ} + \overline{YZ}} \left[\frac{E[\beta_{1kl} \mid \underline{\xi} \in \Omega_{1kl}^m]}{C_t} - \frac{\overline{XYZ}}{\overline{XY} + \overline{XZ} + \overline{YZ}} \right] \right\} \quad (32)$$

where $\overline{X}$, $\overline{Y}$, $\overline{Z}$, $\overline{X^2}$, $\overline{Y^2}$, $\overline{Z^2}$, $\overline{XY}$, $\overline{XZ}$ and $\overline{YZ}$ correspond to $E[X \mid \underline{\xi} \in \Omega_{1kl}^m]$, $E[Y \mid \underline{\xi} \in \Omega_{1kl}^m]$, $E[Z \mid \underline{\xi} \in \Omega_{1kl}^m]$, $E[X^2 \mid \underline{\xi} \in \Omega_{1kl}^m]$, $E[Y^2 \mid \underline{\xi} \in \Omega_{1kl}^m]$, $E[Z^2 \mid \underline{\xi} \in \Omega_{1kl}^m]$, $E[XY \mid \underline{\xi} \in \Omega_{1kl}^m]$, $E[XZ \mid \underline{\xi} \in \Omega_{1kl}^m]$ and $E[YZ \mid \underline{\xi} \in \Omega_{1kl}^m]$, respectively.

Following a similar calculation scheme as in previous sections, the above mentioned $E[.]$ terms can be expressed as a summation of all possible combinations of $E[X_i]$, $E[X_i^2]$, $E[X_iX_j]$, $E[Y_i]$, $E[Y_i^2]$, $E[Y_iY_j]$, $E[Z_i]$, $E[Z_i^2]$, $E[Z_iZ_j]$, $E[X_iY_j]$, $E[X_iZ_j]$ and $E[Y_iZ_j]$ terms.

In details, $E[X_i]$ and $E[Y_i]$ terms are given by the following equations:

$$E[X_i] = \frac{C_{1,i}}{P(\Omega_{1kl}^m)} \int_{-\infty}^{+\infty} 10^{\frac{b\xi_1}{10}} f_{\xi_1}(\xi_1) d\xi_1 \times \begin{cases} \int_{b_k(\xi_1)}^{a_k(\xi_1)} f_{\xi_k}(\xi_k) d\xi_k \int_{b_l(\xi_k)}^{a_l(\xi_k)} f_{\xi_l}(\xi_l) A(a_i(\xi_l),1) \Pi_{i,k,l}(\xi_1) d\xi_l, i \neq k,l \\ \int_{b_k(\xi_1)}^{a_k(\xi_1)} 10^{\frac{b\xi_k}{10}} f_{\xi_k}(\xi_k) d\xi_k \int_{b_l(\xi_k)}^{a_l(\xi_k)} f_{\xi_l}(\xi_l) \Pi_{k,l}(\xi_1) d\xi_l, i = k \\ \int_{b_k(\xi_1)}^{a_k(\xi_1)} f_{\xi_k}(\xi_k) d\xi_k \int_{b_l(\xi_k)}^{a_l(\xi_k)} 10^{\frac{b\xi_l}{10}} f_{\xi_l}(\xi_l) \Pi_{k,l}(\xi_1) d\xi_l, i = l \end{cases} \quad (33)$$

$$E[Y_i] = \frac{C_{k,i}}{P(\Omega_{1kl}^m)} \times \begin{cases} \int_{-\infty}^{+\infty} f_{\xi_1}(\xi_1) d\xi_1 \int_{b_k(\xi_1)}^{a_k(\xi_1)} 10^{-\frac{b\xi_k}{10}} f_{\xi_k}(\xi_k) d\xi_k \int_{b_l(\xi_k)}^{a_l(\xi_k)} f_{\xi_l}(\xi_l) A(a_i(\xi_l),1) \Pi_{i,k,l}(\xi_1) d\xi_l, i \neq 1,l \\ \int_{-\infty}^{+\infty} 10^{\frac{b\xi_1}{10}} f_{\xi_1}(\xi_1) d\xi_1 \int_{b_k(\xi_1)}^{a_k(\xi_1)} 10^{-\frac{b\xi_k}{10}} f_{\xi_k}(\xi_k) d\xi_k \int_{b_l(\xi_k)}^{a_l(\xi_k)} f_{\xi_l}(\xi_l) \Pi_{k,l}(\xi_1) d\xi_l, i = 1 \\ \int_{-\infty}^{+\infty} f_{\xi_1}(\xi_1) d\xi_1 \int_{b_k(\xi_1)}^{a_k(\xi_1)} 10^{-\frac{b\xi_k}{10}} f_{\xi_k}(\xi_k) d\xi_k \int_{b_l(\xi_k)}^{a_l(\xi_k)} 10^{\frac{b\xi_l}{10}} f_{\xi_l}(\xi_l) \Pi_{k,l}(\xi_1) d\xi_l, i = l \end{cases} \quad (34)$$

where the various integration limits in eqs. (33), (34) and throughout this subsection are the same as the ones described in eq. (19). Apparently, the $E[Z_i]$ expressions are similar to the ones in eq. (34):

$$
E[Z_i] = \frac{C_{l,i}}{P(\Omega_{1kl}^m)} \times
\begin{cases}
\displaystyle\int_{-\infty}^{+\infty} f_{\xi_1}(\xi_1)d\xi_1 \int_{b_k(\xi_1)}^{a_k(\xi_1)} f_{\xi_k}(\xi_k)d\xi_k \int_{b_l(\xi_k)}^{a_l(\xi_k)} 10^{-\frac{b\xi_l}{10}} f_{\xi_l}(\xi_l)A(a_i(\xi_l),1)\Pi_{i,k,l}(\xi_l)d\xi_l, i\neq 1,k \\[2ex]
\displaystyle\int_{-\infty}^{+\infty} 10^{\frac{b\xi_1}{10}} f_{\xi_1}(\xi_1)d\xi_1 \int_{b_k(\xi_1)}^{a_k(\xi_1)} f_{\xi_k}(\xi_k)d\xi_k \int_{b_l(\xi_k)}^{a_l(\xi_k)} 10^{-\frac{b\xi_l}{10}} f_{\xi_l}(\xi_l)\Pi_{k,l}(\xi_l)d\xi_l, i=1 \\[2ex]
\displaystyle\int_{-\infty}^{+\infty} f_{\xi_1}(\xi_1)d\xi_1 \int_{b_k(\xi_1)}^{a_k(\xi_1)} 10^{\frac{b\xi_k}{10}} f_{\xi_k}(\xi_k)d\xi_k \int_{b_l(\xi_k)}^{a_l(\xi_k)} 10^{-\frac{b\xi_l}{10}} f_{\xi_l}(\xi_l)\Pi_{k,l}(\xi_l)d\xi_l, i=k
\end{cases}
\tag{35}
$$

$E[X_i^2]$, $E[Y_i^2]$ and $E[Z_i^2]$ expressions can be obtained from eqs. (33)-(35) if we replace $C_{1,i}$ with $C_{1,i}^2$, $C_{k,i}$, with $C_{k,i}^2$, $C_{l,i}$, with $C_{l,i}^2$, $A(a_i,1)$ with $A(a_i,2)$ and $10^{\pm b\xi_1/10}, 10^{\pm b\xi_k/10}$, $10^{\pm b\xi_l/10}$ with $10^{\pm b\xi_1/5}, 10^{\pm b\xi_k/5}$, $10^{\pm b\xi_l/5}$. As in previous cases, $E[X_0]=E[Y_0]=E[Z_0]=(1-u)/P(\Omega_{1kl}^m)$, $E[X_0^2]=E[Y_0^2]=E[Z_0^2](1-u)^2/P(\Omega_{1kl}^m)$.

The $E[X_iX_j]$, $E[Y_iY_j]$ and $E[Z_iZ_j]$ terms can be obtained from eq. (33), (34) and (35) if $10^{-b\xi_1/10}$, $10^{-b\xi_k/10}$ and $10^{-b\xi_l/10}$ is replaced by $10^{-b\xi_1/5}$, $10^{-b\xi_k/5}$ and $10^{-b\xi_l/5}$, respectively. In addition, if I and $j\neq k$ and l in eq. (33), I and $j\neq 1$ and l in eq. (34) and I and $j\neq 1$ and k in eq. (35),

$A(a_i(\xi_l),1)\Pi_{i,k,l}(\xi_l)$ should be replaced by $P(\xi_l) = A(a_i(\xi_l),1)A(a_j(\xi_l),1) \displaystyle\prod_{n=2,n\neq i,j,k,l}^{19} A(a_n(\xi_l),0)$.

Furthermore, if $i=k$ or l in eq. (33), $i=1$ or l in eq. (34) and $i=1$ or k in eq. (35), then $\Pi_{k,l}(\xi_l)$ should be replaced by $A(a_j(\xi_l),1)\Pi_{j,k,l}(\xi_l)$. Finally, if j takes the latter I values, the same expressions still apply if we interchange I with j.

The cross product terms $E[X_iY_j]$, $E[X_iZ_j]$ and $E[Y_iZ_j]$ are expressed below:

$$
E[X_iY_j] = \frac{C_{1,i}C_{k,j}}{P(\Omega_{1kl}^m)} \int_{-\infty}^{+\infty} 10^{-\frac{b\xi_1}{10}} f_{\xi_1}(\xi_1)d\xi_1 \int_{b_k(\xi_1)}^{a_k(\xi_1)} 10^{-\frac{b\xi_k}{10}} f_{\xi_k}(\xi_k)d\xi_k \int_{b_l(\xi_k)}^{a_l(\xi_k)} f_{\xi_l}(\xi_l)P(\xi_l)d\xi_l
$$

(36)

$$
E[X_iZ_j] = \frac{C_{1,i}C_{l,j}}{P(\Omega_{1kl}^m)} \int_{-\infty}^{+\infty} 10^{-\frac{b\xi_1}{10}} f_{\xi_1}(\xi_1)d\xi_1 \int_{b_k(\xi_1)}^{a_k(\xi_1)} f_{\xi_k}(\xi_k)d\xi_k \int_{b_l(\xi_k)}^{a_l(\xi_k)} 10^{-\frac{b\xi_l}{10}} f_{\xi_l}(\xi_l)P(\xi_l)d\xi_l \tag{37}
$$

$$
E[Y_iZ_j] = \frac{C_{k,i}C_{l,j}}{P(\Omega_{1kl}^m)} \int_{-\infty}^{+\infty} f_{\xi_1}(\xi_1)d\xi_1 \int_{b_k(\xi_1)}^{a_k(\xi_1)} 10^{-\frac{b\xi_k}{10}} f_{\xi_k}(\xi_k)d\xi_k \int_{b_l(\xi_k)}^{a_l(\xi_k)} 10^{-\frac{b\xi_l}{10}} f_{\xi_l}(\xi_l)P(\xi_l)d\xi_l \tag{38}
$$

where we assume $i\neq k$, l, $j\neq 1$, l and $i\neq j$ in eq. (36), $i\neq k$, l, $j\neq 1$, k and $i\neq j$ in eq. (37) and $i\neq 1$, l, $j\neq 1$, k and $i\neq j$ in eq. (38). If $i=k$ or l the $10^{b\xi_i/10}$ term is transferred to the kth integral (thus

$10^{-b\xi_k/10}$ vanishes in eq. (36)) or to the lth integral (thus $10^{-b\xi_l/10}$ vanishes in eq. (37)-(38)). In addition, if $i=1$ in eq. (38) $10^{b\xi_i/10}$ term is transferred to the 1st integral. In all aforementioned cases $P(\xi_l)$ converts to $A(a_j(\xi_l),1)\Pi_{j,k,l}(\xi_l)$. If $j=1$ or l the $10^{b\xi_j/10}$ term is transferred to the 1st integral (thus $10^{-b\xi_1/10}$ vanishes in eqs. (36)-(37)) or to the lth integral. In addition, if $j=k$ in eqs. (37)-(38) the $10^{b\xi_j/10}$ term is transferred to the kth integral (thus $10^{-b\xi_k/10}$ vanishes in eq. (38)). In all aforementioned cases $P(\xi_l)$ converts to $A(a_i(\xi_l),1)\Pi_{i,k,l}(\xi_l)$. Finally, if $i=j\neq l$ in eq. (36), $i=j\neq k$ in eq. (37) and $i=j\neq1$ in eq. (38) then $P(\xi_l)$ converts to $A(a_j(\xi_l),2)\Pi_{j,k,l}(\xi_l)$. Otherwise, if $i=j=l$ in eq. (36), $i=j=k$ in eq. (37) and $i=j=1$ in eq. (38) then the $10^{b\xi_l/5}$, $10^{b\xi_k/5}$ and $10^{b\xi_1/5}$ term appears in the lth, k^{th} and 1st integral, respectively, whereas $P(\xi_l)$ converts to $\Pi_{k,l}(\xi_l)$.

V. NUMERICAL RESULTS & DISCUSSION

First, a comparison between the calculations of the proposed theoretical model and the corresponding ones from an independent numerical simulation will be discussed. The calculations have been performed with respect to the expected value $E[\beta| \underline{\xi} \in \Omega^m]$ $(=\overline{\beta})$ and the standard deviation $\sqrt{E[\beta^2 | \underline{\xi} \in \Omega^m]-\left(E[\beta | \underline{\xi} \in \Omega^m]\right)^2}$ $(=\sigma_\beta)$ of power consumption. The under examination scenarios include different MS positions (r,θ), various path loss factors (α) and standard deviations of shadowing losses (σ), as well as different AS sizes, cell selection thresholds (cst) and SHO thresholds (sht). The service parameters correspond to a typical voice service in WCDMA UMTS networks: $v=0.5$, $R=12.2$ Kbps, $W=3.84$ Mchips/s and $[E_b/I_o]_t=4.4$ dB. Finally, the orthogonality factor is $u=0.9$.

The numerical simulation model has been configured to generate 100.000 random shadowing samples according to a log-normal pdf. For each sample the cell selection and the handoff inequalities of Section III are examined, first to decide whether the sample refers to the cell under examination or not and next to decide which of the three handoff conditions is fulfilled. According to the latter criterion a power consumption sample is calculated using one of the equations (2)-(4), and next $\overline{\beta}$ and σ_β is estimated using equations (13) and (14), respectively. In order to facilitate a tabulated comparison between the numerical results and the corresponding theoretical ones the results from 5 rounds of simulation runs have been averaged and presented in Tables I, II and III. Each Table refers to a different scenario and proves that theoretical and numerical estimations converge, which in turn proves the efficiency of the Taylor series approximation.

Next, in order to demonstrate the potential benefits from the adaptation of the proposed theoretical model the power consumption statistics will be further investigated. The under examination numerical results are illustrated in Figs. 3-8. Figs. 3, 5 and 7 depict $\overline{\beta}$ for $AS=1$, 2 and 3, respectively, versus the normalized distance r_1/R_{max}. Figs 4, 6 and 8 depict σ_β for the former scenarios.

Fig. 3 corresponds to a HHO scenario. According to the illustrated data $\overline{\beta}$ tends to increase, as expected, when the MS approaches the cell border. Near BS and up to a distance, $\overline{\beta}$ increases, when α and σ take higher values. Nevertheless, this is not valid, when the MS

approaches the cell border. Actually, close to the border a hostile propagation environment (i.e. high α and σ values) results to less power consumption. This behavior can be explained, if we take into the account the possibility of handoff. Close to the border the *MS* tends to camp to another cell instead of sustaining the degradation of a hostile environment. Actually, this is more evident, when the cell selection criterion is more tight, i.e. *cst*=1 instead of *cst*=3, and camping to another cell is encouraged. The comments from Fig. 4 are rather similar to the ones in Fig. 3. The higher (lower) σ_β appears, when α and σ take lower (higher) and the cell selection algorithm decision criteria are relatively loose (tight). According to the aforementioned comments the cell selection imperfections burdens the system, when the propagation conditions are relatively good and *AS*=1. In such cases, the *MS* should be encouraged to camp to a neighbor cell.

Fig. 5 illustrates the expected value of power consumption, when *AS*=2 and thus a 2-way *SHO* is also possible. Fig. 5 also includes results for *AS*=1 for comparison reasons. According to the illustrated data the highest values of $\overline{\beta}$ appear, when α and σ take low values as it was already mentioned in Fig.3. If we compare *AS*=1 and *AS*=2 results, it appears that the choice of *AS*=2 and more than this the encouragement of *SHO* is beneficial and this is more evident when the *MS* approaches the cell border. Actually, when α and σ take low values and the *MS* moves towards the cell border/corner *SHO* takes advantage of the good propagation conditions and allows one neighbor *BS* to participate instead of being a strong interferer. Fig. 6 illustrates σ_β numerical results for the network scenarios examined in Fig. 5. According to the illustrated results, the option and more than this the encouragement of *SHO* reduces significantly σ_β at least when compared to *AS*=1 scenarios. Concluding, the inclusion of a *SHO* option by setting *AS*=2, provides significant benefits, in terms of reducing $\overline{\beta}$ and σ_β, even in cases where the MS is located relatively close to the BS.

Fig. 7 illustrates the expected value of power consumption, when *AS*=3 and thus a 3-way *SHO* is also possible. According to the illustrated data the *AS*=3 choice gives slightly better results, when is compared with the relevant results of Fig. 5 and particular with the case of σ=8 dB. However, the encouragement of *SHO* (*sht*=3 dB) provides a significant reduction, when compared with the *AS*=1 and *AS*=2 choice and the case of σ=10 dB. Fig. 8 illustrates σ_β for the network scenarios examined in Fig. 7. According to the illustrated results and the comparison with the relevant results in Fig.6, the choice of *AS*=3 and the encouragement of *SHO* provides a significant reduction of σ_β and a location insensitive behavior.

Concluding the discussion on the aforementioned results it is worthwhile to mention that as it has been found in similar research works the resource allocation on CDMA networks strongly depends on the propagation conditions, the MS location and the various Radio Resource Management (RRM) settings. Thus, an optimized network performance definitely requires a cross layer approach and prediction models that can incorporate both physical layer and RRM parameters.

VI. CONCLUSION

A theoretical statistical model that provides an estimation of the expected and standard deviation value of power consumption in the downlink direction has been developed for cellular CDMA networks. The proposed model supports the aforementioned calculations taking into account cell selection and handoff settings. In this context, present work contributes to a cross-layer approach, by establishing a theoretical framework, which facilitates performance evaluation and optimization of CDMA networks under specific radio propagation conditions as well as RRM settings. Current work can be extended with future studies in several directions. The most challenging future extension is to provide a joint pdf for power consumption based on the capability to estimate power consumption moments. Furthermore, present work provides estimations on a link level and thus an extension of the model in order to support performance evaluation on a network level is also another interesting research direction. A cross layer design approach aiming to develop an optimized soft handoff algorithm, which will take into account the proposed model's estimations, is another one possible future research topic. Finally, the under consideration numerical results are based on several assumptions, which can be easily rearranged. For example, it would be interesting to produce numerical results by taking into account unbalanced power allocation schemes among the *SHO* links or unequal traffic loads per cell.

APPENDIX I

In the case of 2-way SHO connections, β_{1k} power consumption metric can be expressed in the form of the following function:

$$\beta(X,Y) = \left(\frac{1}{X}+\frac{1}{Y}\right)^{-1} = \frac{XY}{X+Y} \tag{I.1}$$

Using a Taylor expansion in the neighborhood of $E[X(\xi)]=\overline{X}$ and $E[Y(\xi)]=\overline{Y}$, where Taylor series terms higher than the second order are omitted, and next taking the average value of this expression we conclude after a few straightforward calculations to:

$$E[\beta(X,Y)] = \beta(\overline{X},\overline{Y}) + E\left[(X-\overline{X})^2\right]\frac{\partial^2}{\partial X^2}[\beta(X,Y)]_{\substack{X=\overline{X}\\Y=\overline{Y}}} + E\left[(Y-\overline{Y})^2\right]\frac{\partial^2}{\partial Y^2}[\beta(X,Y)]_{\substack{X=\overline{X}\\Y=\overline{Y}}} +$$

$$+2E\left[(X-\overline{X})(Y-\overline{Y})\right]\frac{\partial^2}{\partial X\partial Y}[\beta(X,Y)]_{\substack{X=\overline{X}\\Y=\overline{Y}}} \tag{I.2}$$

where

$$\frac{\partial^2}{\partial X^2}[\beta(X,Y)]_{\substack{X=\overline{X}\\Y=\overline{Y}}} = -\frac{2(\overline{Y})^2}{(\overline{X}+\overline{Y})^3}, \frac{\partial^2}{\partial Y^2}[\beta(X,Y)]_{\substack{X=\overline{X}\\Y=\overline{Y}}} = -\frac{2(\overline{X})^2}{(\overline{X}+\overline{Y})^3}, \frac{\partial^2}{\partial X\partial Y}[\beta(X,Y)]_{\substack{X=\overline{X}\\Y=\overline{Y}}} = \frac{2\overline{XY}}{(\overline{X}+\overline{Y})^3} \tag{I.3}$$

By taking the square power of the above mentioned Taylor series expansion and omitting higher order terms, we conclude, after some manipulation, to the following expression regarding $E\left[\beta^2(X,Y)\right]$:

$$E\left[\beta^2(X,Y)\right] = \beta^2(\overline{X},\overline{Y}) + E\left[(X-\overline{X})^2\right]\left(\frac{\partial}{\partial X}[\beta(X,Y)]_{\substack{X=\overline{X}\\Y=\overline{Y}}}\right)^2 + E\left[(Y-\overline{Y})^2\right]\left(\frac{\partial}{\partial Y}[\beta(X,Y)]_{\substack{X=\overline{X}\\Y=\overline{Y}}}\right)^2 +$$

$$+2E\left[(X-\overline{X})(Y-\overline{Y})\right]\left(\frac{\partial}{\partial X}[\beta(X,Y)]_{\substack{X=\overline{X}\\Y=\overline{Y}}}\right)\left(\frac{\partial}{\partial Y}[\beta(X,Y)]_{\substack{X=\overline{X}\\Y=\overline{Y}}}\right) + 2\beta(\overline{X},\overline{Y})\{E[\beta(X,Y)]-\beta(\overline{X},\overline{Y})\} \tag{I.4}$$

where

$$\frac{\partial}{\partial X}[\beta(X,Y)]_{\substack{X=\overline{X}\\Y=\overline{Y}}} = \frac{(\overline{Y})^2}{(\overline{X}+\overline{Y})^2}, \frac{\partial}{\partial Y}[\beta(X,Y)]_{\substack{X=\overline{X}\\Y=\overline{Y}}} = \frac{(\overline{X})^2}{(\overline{X}+\overline{Y})^2} \tag{I.5}$$

APPENDIX II

In the case of 3-way SHO connections, β_{1kl} power consumption metric can be expressed in the form of the following function:

$$\beta(X,Y,Z) = \left(\frac{1}{X}+\frac{1}{Y}+\frac{1}{Z}\right)^{-1} = \frac{XYZ}{XY+XZ+YZ} \tag{II.1}$$

Using a Taylor expansion as in Appendix II, we conclude after a few straightforward calculations to:

$$E[\beta(X,Y,Z)] = \beta(\overline{X},\overline{Y},\overline{Z}) + E[(X-\overline{X})^2]\frac{\partial^2}{\partial X^2}[\beta(X,Y,Z)]_{\substack{X=\overline{X}\\Y=\overline{Y}\\Z=\overline{Z}}} + E[(Y-\overline{Y})^2]\frac{\partial^2}{\partial Y^2}[\beta(X,Y,Z)]_{\substack{X=\overline{X}\\Y=\overline{Y}\\Z=\overline{Z}}} +$$

$$+ E[(Z-\overline{Z})^2]\frac{\partial^2}{\partial Z^2}[\beta(X,Y,Z)]_{\substack{X=\overline{X}\\Y=\overline{Y}\\Z=\overline{Z}}} + 2E[(X-\overline{X})(Y-\overline{Y})]\frac{\partial^2}{\partial X\partial Y}[\beta(X,Y,Z)]_{\substack{X=\overline{X}\\Y=\overline{Y}\\Z=\overline{Z}}} +$$

$$+ 2E[(X-\overline{X})(Z-\overline{Z})]\frac{\partial^2}{\partial X\partial Z}[\beta(X,Y,Z)]_{\substack{X=\overline{X}\\Y=\overline{Y}\\Z=\overline{Z}}} + 2E[(Y-\overline{Y})(Z-\overline{Z})]\frac{\partial^2}{\partial Y\partial Z}[\beta(X,Y,Z)]_{\substack{X=\overline{X}\\Y=\overline{Y}\\Z=\overline{Z}}} \tag{II.2}$$

where

$$\frac{\partial^2}{\partial X^2}[\beta(X,Y,Z)]_{\substack{X=\overline{X}\\Y=\overline{Y}\\Z=\overline{Z}}} = -\frac{2(\overline{YZ})^2(\overline{Y}+\overline{Z})}{(\overline{XY}+\overline{XZ}+\overline{YZ})^3}, \frac{\partial^2}{\partial Y^2}[\beta(X,Y,Z)]_{\substack{X=\overline{X}\\Y=\overline{Y}\\Z=\overline{Z}}} = -\frac{2(\overline{XZ})^2(\overline{X}+\overline{Z})}{(\overline{XY}+\overline{XZ}+\overline{YZ})^3},$$

$$\frac{\partial^2}{\partial Z^2}[\beta(X,Y,Z)]_{\substack{X=\overline{X}\\Y=\overline{Y}\\Z=\overline{Z}}} = -\frac{2(\overline{XY})^2(\overline{X}+\overline{Y})}{(\overline{XY}+\overline{XZ}+\overline{YZ})^3} \tag{II.3}$$

$$\frac{\partial^2}{\partial X\partial Y}[\beta(X,Y,Z)]_{\substack{X=\overline{X}\\Y=\overline{Y}\\Z=\overline{Z}}} = \frac{2\overline{XY}(\overline{Z})^3}{(\overline{XY}+\overline{XZ}+\overline{YZ})^3}, \frac{\partial^2}{\partial X\partial Z}[\beta(X,Y,Z)]_{\substack{X=\overline{X}\\Y=\overline{Y}\\Z=\overline{Z}}} = \frac{2\overline{X}(\overline{Y})^3\overline{Z}}{(\overline{XY}+\overline{XZ}+\overline{YZ})^3}$$

$$\frac{\partial^2}{\partial Y\partial Z}[\beta(X,Y,Z)]_{\substack{X=\overline{X}\\Y=\overline{Y}\\Z=\overline{Z}}} = \frac{2(\overline{X})^3\overline{YZ}}{(\overline{XY}+\overline{XZ}+\overline{YZ})^3} \tag{II.4}$$

Using the square power of the Taylor series expansions and omitting higher order terms, we conclude to the following expression regarding $E[\beta^2(X,Y,Z)]$:

$$E\left[\beta^2(X,Y,Z)\right] = \beta^2(\overline{X},\overline{Y},\overline{Z}) + E\left[(X-\overline{X})^2\right]\left(\frac{\partial}{\partial X}\left[\beta(X,Y,Z)\right]_{\substack{X=\overline{X}\\Y=\overline{Y}\\Z=\overline{Z}}}\right)^2 +$$

$$+ E\left[(Y-\overline{Y})^2\right]\left(\frac{\partial}{\partial Y}\left[\beta(X,Y,Z)\right]_{\substack{X=\overline{X}\\Y=\overline{Y}\\Z=\overline{Z}}}\right)^2 + E\left[(Z-\overline{Z})^2\right]\left(\frac{\partial}{\partial Z}\left[\beta(X,Y,Z)\right]_{\substack{X=\overline{X}\\Y=\overline{Y}\\Z=\overline{Z}}}\right)^2 +$$

$$+ 2E\left[(X-\overline{X})(Y-\overline{Y})\right]\left(\frac{\partial}{\partial X}\left[\beta(X,Y,Z)\right]_{\substack{X=\overline{X}\\Y=\overline{Y}\\Z=\overline{Z}}}\right)\left(\frac{\partial}{\partial Y}\left[\beta(X,Y,Z)\right]_{\substack{X=\overline{X}\\Y=\overline{Y}\\Z=\overline{Z}}}\right) +$$

$$+ 2E\left[(X-\overline{X})(Z-\overline{Z})\right]\left(\frac{\partial}{\partial X}\left[\beta(X,Y,Z)\right]_{\substack{X=\overline{X}\\Y=\overline{Y}\\Z=\overline{Z}}}\right)\left(\frac{\partial}{\partial Z}\left[\beta(X,Y,Z)\right]_{\substack{X=\overline{X}\\Y=\overline{Y}\\Z=\overline{Z}}}\right) +$$

$$+ 2E\left[(Y-\overline{Y})(Z-\overline{Z})\right]\left(\frac{\partial}{\partial Y}\left[\beta(X,Y,Z)\right]_{\substack{X=\overline{X}\\Y=\overline{Y}\\Z=\overline{Z}}}\right)\left(\frac{\partial}{\partial Z}\left[\beta(X,Y,Z)\right]_{\substack{X=\overline{X}\\Y=\overline{Y}\\Z=\overline{Z}}}\right) +$$

$$+ 2\beta(\overline{X},\overline{Y},\overline{Z})\left\{E\left[\beta(X,Y,Z)\right] - \beta(\overline{X},\overline{Y},\overline{Z})\right\} \tag{II.5}$$

where

$$\frac{\partial}{\partial X}\left[\beta(X,Y,Z)\right]_{\substack{X=\overline{X}\\Y=\overline{Y}\\Z=\overline{Z}}} = \frac{\left(\overline{Y}\,\overline{Z}\right)^2}{\left(\overline{X}\,\overline{Y} + \overline{X}\,\overline{Z} + \overline{Y}\,\overline{Z}\right)^2}, \frac{\partial}{\partial Y}\left[\beta(X,Y,Z)\right]_{\substack{X=\overline{X}\\Y=\overline{Y}\\Z=\overline{Z}}} = \frac{\left(\overline{X}\,\overline{Z}\right)^2}{\left(\overline{X}\,\overline{Y} + \overline{X}\,\overline{Z} + \overline{Y}\,\overline{Z}\right)^2},$$

$$\frac{\partial}{\partial Z}\left[\beta(X,Y,Z)\right]_{\substack{X=\overline{X}\\Y=\overline{Y}\\Z=\overline{Z}}} = \frac{\left(\overline{X}\,\overline{Y}\right)^2}{\left(\overline{X}\,\overline{Y} + \overline{X}\,\overline{Z} + \overline{Y}\,\overline{Z}\right)^2} \tag{II.6}$$

REFERENCES

1. K. S.Gilhousen, I. M. Jacobs, R Padovani, A. J. Viterbi, L. A. Weaver and C. E. Wheatley, "On the capacity of a cellular CDMA system," *IEEE Trans. Veh. Technol.*, vol. 40, pp. 303-312, May 1991.
2. J. Viterbi, A.M.Viterbi and K.S. Gilhousen, "Soft Handoff extends CDMA coverage and increase reverse link capacity," *IEEE J. Select. Areas Commun.*, Vol. 4, pp. 1281-1288, Oct. 1994.
3. M. Zorzi, "On the analytical computation of the interference statistics with applications to the performance of mobile radio systems," *IEEE Trans. Commun.*, vol. 45, pp. 103-109, Jan. 1997.
4. V. V. Veeravalli and A. Sendonaris, "The coverage-capacity tradeoff in cellular CDMA systems," *IEEE Trans. Veh. Technol.*, vol. 48, pp. 1443-1450, Sept. 1999.
5. R. Pillay and F. Takawira, "Performance analysis of soft handoff in CDMA cellular networks," *IEEE Trans. Veh. Technol.*, vol. 50, pp. 1507-1517, Nov. 2001.
6. H. Jiang and C.H. Davis, "Coverage expansion and capacity improvement from soft handoff for CDMA cellular systems", *IEEE Trans. Wireless Commun.*, vol. 4, pp. 2163-2171, Sep. 2005.
7. F. Adeltando, J.Perez-Romero and O. Sallent "An Analytical model for the reverse link of WCDMA systems with repeaters in nonuniform traffic distributions", *IEEE Trans. Veh. Technol.*, vol. 58, pp. 2180-2190, Jun. 2009.

8. H. Jo, C. Mun, J. Moon and J. Yook, "Interference mitigation using uplink control for two-tier femtocell networks", *IEEE Trans. Wireless Commun.*, vol. 8, pp. 4906-4910, Oct. 2009.

9. C. Lee and R. Steele, "Effect of soft and softer handoffs on CDMA system capacity," *IEEE Trans. Veh. Technol.*, vol. 47, pp. 830-841, Aug. 1998.

10. M. Pratesi, F. Santucci, F. Graziosi and M. Ruggieri, "Outage analysis in mobile radio systems with generically correlated log-normal interferers," *IEEE Trans. Commun.*, vol. 48, pp. 381-385, Mar. 2000.

11. W. Choi and J. Y. Kim, "Forward-Link Capacity of a DS/CDMA system with mixed multirate sources," *IEEE Trans. Veh. Technol.*, vol. 50, pp. 737-749, May 2001.

12. C. Jaeweon, and H. Daehyoung, "Statistical model of downlink interference for the performance evaluation of CDMA Systems," *IEEE Commun. Lett.*, vol. 6, pp. 494-496, Nov. 2002.

13. Y. Chen, and L. Cuthbert, "Optimum size of soft handover zone in power-controlled UMTS downlink systems," *IEEE Electronics Lett.*, vol. 38, pp 89 -90, Jan. 2002.

14. Y. Chen and L. Cuthbert, "Downlink radio resource optimization in wide-band CDMA systems", *Wirel. Commun. Mob. Comput.*, vol. 3, pp. 735-742, Nov. 2003.

15. D. Avidor, N. Hegde and S. Mukherjee, "On the impact of the soft handoff threshold and the maximum size of the active group on resource allocation and outage probability in the UMTS system", *IEEE Trans. Wireless Commun.*, vol. 3, pp. 565-577, Mar. 2004.

16. D. Zhao, X. Shen and J. W. Mark, "Soft handoff and connection reliability in cellular CDMA downlinks", *IEEE Trans. Wireless Commun.*, vol. 5, pp. 354-365, Feb. 2006.

17. J. Koo, Y. Han and J. Kim, "Handoff effect on CDMA forward link capacity", *IEEE Trans. Wireless Commun.*, vol. 5, pp. 262-269, Feb. 2006.

18. D. Li and V.K. Prabhu, "Effects of the BS power and soft handoff on the outage and capacity in the forward link of an SIR-based power-controlled CDMA system", *IEEE Trans. Wireless Commun.*, vol. 5, pp. 1987-1992, Aug. 2006.

19. Viterbi, A. (1995). CDMA: Principles of Spread Spectrum Communication. Reading, MA: Addison-Wesley, pp. 218-227.

20. S. S. Rizvi, A. Riasat and K. M. Elleithy, "A quantitative analysis of handover time at MAC layer for wireless mobile networks," *Int. J. Wireless & Mobile Networks (IJWMN)*, vol. 1, no 2, Nov. 2009.

21. C-C Lin, K. Sandrasegaran, H. A. M. Ramli and R. Basukala, "Optimized performance evaluation of LTE hard handover algorithm with average RSRP constraint," *Int. J. Wireless & Mobile Networks (IJWMN)*, vol. 3, no 2, April 2011.

Table I. Theoretical vs Numerical Simulation Estimations for AS=1

AS=1 $\theta=15^{o}$, $a=3, \sigma=8, cst=1$		$r=0.6R_{max}$	$r=0.7R_{max}$	$r=0.8R_{max}$	$r=0.9_{max}R$	$r=1.0R_{max}$
Theoretical Model	$\overline{\beta}$	0.0035164	0.0043695	0.0051464	0.0058282	0.0064095
	σ_{β}	0.0032682	0.0037155	0.0040348	0.0042499	0.0043742
Numerical Model	$\overline{\beta}$	0.0035124	0.0043708	0.0051283	0.0057727	0.0064118
	σ_{β}	0.0032512	0.0036875	0.0040272	0.0041949	0.0043510

Table II. Theoretical vs Numerical Simulation Estimations for AS=2

AS=2 $\theta=30^o$, $a=3, \sigma=8, cst=1$, $sht=3$		r=0.6R$_{max}$	r=0.7R$_{max}$	r=0.8R$_{max}$	r=0.9R$_{max}$	r=1.0R$_{max}$
Theoretical Model	$\bar{\beta}$	0.0031774	0.0036789	0.0040796	0.0043946	0.0046417
	σ_β	0.0017504	0.0018292	0.0018618	0.0018691	0.0018788
Numerical Model	$\bar{\beta}$	0.0031854	0.0036904	0.0040845	0.0044148	0.0046605
	σ_β	0.0017671	0.0018521	0.0018910	0.0019045	0.0019193

Table III. Theoretical vs Numerical Simulation Estimations for AS=3

AS=3 $\theta=0^o$, $a=4, \sigma=10, cst=1$, $sht=3$		r=0.6R$_{max}$	r=0.7R$_{max}$	r=0.8R$_{max}$	r=0.9R$_{max}$	r=1.0R$_{max}$
Theoretical Model	$\bar{\beta}$	0.0016295	0.0019904	0.0022905	0.0025476	0.0027399
	σ_β	0.0010969	0.0012158	0.0012493	0.0012537	0.0012392
Numerical Model	$\bar{\beta}$	0.0016487	0.0020319	0.0023516	0.0026368	0.0028613
	σ_β	0.0011215	0.0012532	0.0013089	0.0013289	0.0013308

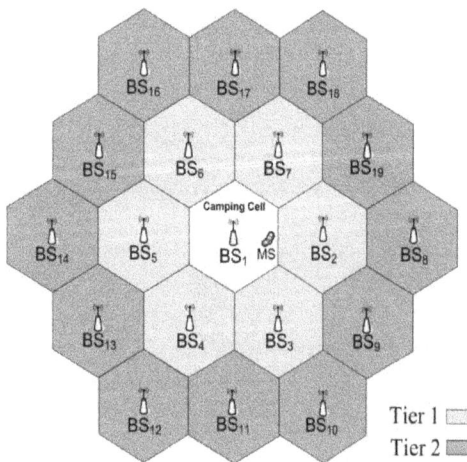

Fig. 1. The considered cellular network

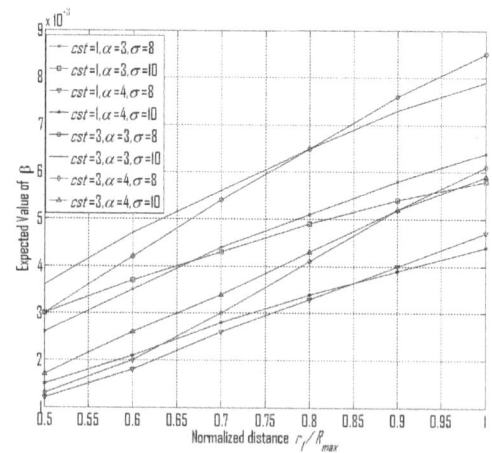

Fig 3. Expected value of power consumption versus normalized distance r_1/R_{max} (AS=1 and $\theta=15^o$)

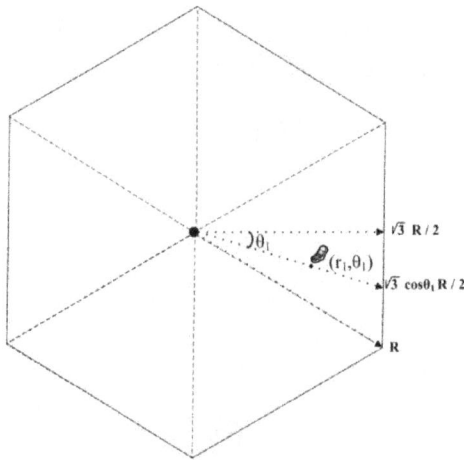

Fig. 2. The cell geometry and the spatial coordinates r, θ.

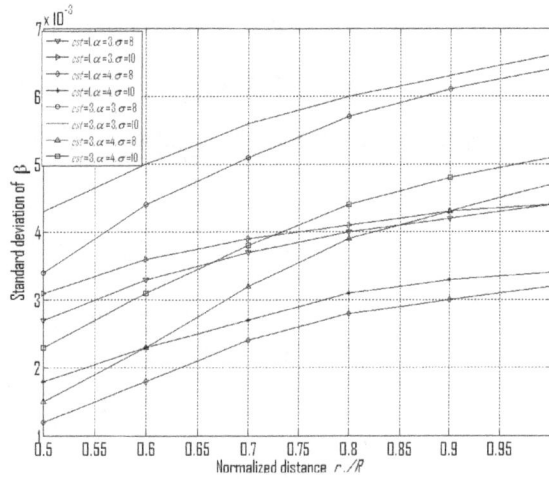

Fig 4. Standard deviation of power consumption versus normalized distance r_1/R_{max} (AS=1 and θ=15°)

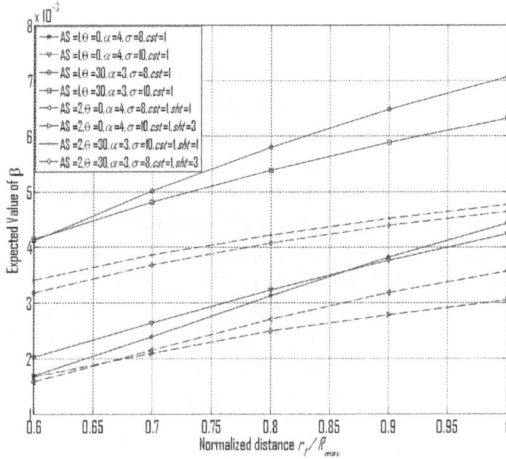

Fig 5. Expected value of power consumption versus normalized distance r_1/R_{max} (AS=1, 2).

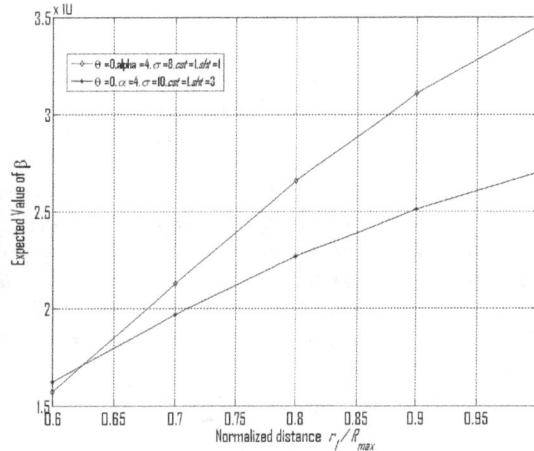

Fig 6. Standard deviation of power consumption versus normalized distance r_1/R_{max} (AS=1. 2).

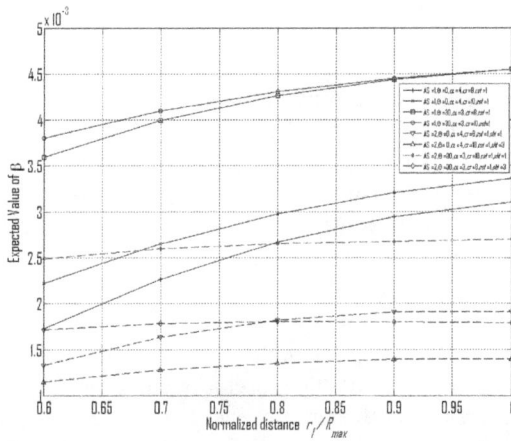

Fig 7. Expected value of power consumption versus normalized distance r_1/R_{max} (AS=3).

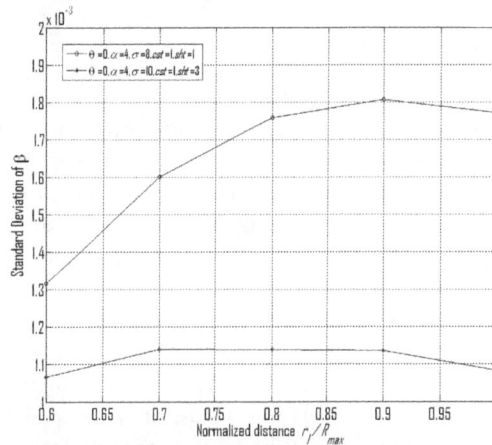

Fig 8. Standard deviation of power consumption versus normalized distance r_1/R_{max} (AS=3).

5

SIMPLIFIED PERFORMANCE ANALYSIS OF ENERGY DETECTORS OVER MYRIAD FADING CHANNELS: AREA UNDER THE ROC CURVE APPROACH

Shumon Alam, O. Olabiyi, O. Odejide, and A. Annamalai

Center of Excellence for Communication Systems Technology Research
Department of Electrical and Computer Engineering,
Prairie View A & M University, TX 77446 United States of America
shalam2000@gmail.com,
engr3os@gmail.com, femiodejide@yahoo.com, aaannamalai@pvamu.edu

ABSTRACT

This article investigates the energy detector's performance in myriad fading environments by exploiting a canonical series representation of the generalized Marcum Q-function (along with higher-order derivatives of the moment generating function (MGF) of end-to-end signal-to-noise ratio (SNR) in closed-form) or by a single integral formula involving the cumulative distribution function (CDF) of the end-to-end SNR. The detection performance is characterized by the area under the receiver operating characteristic (ROC) curve (AUC), which is a simple statistical performance measuring metric that varies between 0.5 and 1. The proposed frameworks are capable of handling the half-odd positive integer values of the time-bandwidth products (u) and the fading severity indices (m) of the Nakagami-m distribution. The methods simplify the cases where many were either intractable or required complicated mathematical steps with the classical probability density function (PDF)/contour integral approaches. Several receiver diversity combining schemes are examined to demonstrate the versatility of the frameworks over independent but not necessarily identically distributed fading channels. Composite effects (multipath fading + shadowing) are also considered for the analysis. Our proposed frameworks with AUC performance metric in the myriad fading environments provide new ways of characterizing and evaluating energy detector's performance.

KEYWORDS

Area under Curve, Diversity Combining, Energy Detector, Receiver Operating Characteristic (ROC)

1. INTRODUCTION

Cognitive radio (CR) holds promise to alleviate today's spectrum scarcity and underutilization problem by seeking transmission opportunities over spectral bands that are temporarily unoccupied or in a non-interfering manner and without primary user negotiation. CRs are designed to sense the environment dynamically and tune their transmission parameters rapidly to best utilize the vacant or underutilized premium spectrum real-estate[1]. Therefore, the first "cognitive" task preceding any form of dynamic spectrum management is to sense the spectrum for identification of the unoccupied white (a subband only occupied by white noise) and gray (occupied by interferers and noise) spaces. Among the various known spectrum sensing techniques, blind sensing based energy detector is perhaps the simplest (i.e., low-complexity) and most versatile since it does not require a priori knowledge of other user transmissions except of the transmission bandwidths and center frequencies[2]. In this article, energy detector's performance is analyzed using the area under the receiver operating characteristic (ROC) curve (AUC)[3],[4]. Although, the ROC curve (probability of detection vs. probability of false alarm) or the complementary ROC curve (probability of missed detection vs. probability of false alarm) is capable of characterizing the energy detector's performance completely, sometimes it is difficult to compare detectors' performance using their ROC curves especially when the curves overlap.Therefore, it is desirable to have a single figure of merit that can give distinction between two detectors. Such a measure is the area under the ROC curve (AUC) which varies between 0.5 (poor performance) and 1 (good performance)[5].It is a single scalar value of expected performance and represents the probability that choosing the correct decision at the detector is more likely than choosing the incorrect decision[4]. In the worst case scenario, practical detector's performance can be described by AUC of 0.5but the value increases to 1 as the detection improves. Even though, AUC is a useful parameter for describing detector's performance, it is only recently addressed in[5].

The research community agrees that the detection of white/gray space is more tractable by means of detection of unknown deterministic signals by energy detectors. Urkowitz [6] studied first the detection of an unknown deterministic signal over the Gaussian channel in the presence of noise through an energy detector. The presence of the energy is detected by comparing the energy of the received signal over a period with the predefined energy threshold (λ). This method has been extended to various scenarios and diversity combining schemes. It is important to note that most detection analyses in the literature are based on the probability density function (PDF) and the analyses use the ROC curves to describe the performance (e.g., [7]-[9]). The diversity combining isconsidered solely either over independent and identically distributed (*i.i.d.*) Rayleigh or Nakagami-m (restricted to positive integer fading index m) channels due to the lack of closed-form solutions for integrals involving the generalized Marcum Q-function. Moreover, the analyses are restricted to integer time-bandwidth product (TW) / sampling number (u). Although Rice fading was considered in[7], the result is limited to the unity value of u. Contour integral is considered in[8] for correlated dual diversity for Rice channels but it is also restricted to integer u value. An exponential-type contour integral is used in[9] to represent the generalized Marcum Q-function to provide a closed-form solution over *i.i.d.* Nakagami-*m* fading with positive integer mL (where L denotes the diversity order) for maximum ratio combining (MRC) detector. [9]also provides results for *i.i.d.* Rice channels using the Laurent series expansion of exponential term but the final solution is in the form of an infinite series; the summation terms require the evaluation of two separate higher order derivatives of a product term [9, eq. (27)]. Atapattu et al. [5]characterized energy detection-performance using AUC only over Nakagami-m channel and analyzed diversity combing techniques for the *i.i.d.* cases (also restricted to integer m and u) but the framework itself is complicated with multiple summations terms and the regularized confluent hypergeometric function of the confluent hypergeometric function $_2F_1(.\ ;\ .\ ;\ .)$. Moreover, it cannot be generalized easily for other fading channels. Thus, most of the prior works have limitations; their solutions are not easy to generalize for myriad fading environments and intractable for the independent, non-identically distributed (*i.n.d.*) diversity combining schemes.

Thus, motivated by the current limitations, we seek simple alternative approaches for analyzing the detection-performance over myriad *i.i.d./i.n.d.* channel conditions.To overcome the mathematical limitations, we develop two analytical approaches to characterize the AUC performance metric over various propagation channel models (including multipath and composite effects) either by exploiting a canonical infinite series representation of the generalized Marcum Q-function (along with higher-order derivatives of the moment generating function (MGF) of end-to-end SNR in closed-form) or by a single integral formula involving the cumulative distribution function (CDF) of the end-to-end SNR. The MGF approach has advantages over other approaches and can be seen to be utilized in many recent works(see e.g.,[10]).We use the proposed frameworks subsequently to study the efficacies of diversity combining schemes such as the maximum ratio combining (MRC), square law combining (SLC), selection combining (SC) and the switch and stay combining (SSC) schemes over *i.i.d./i.n.d.* channels subject to multipath fading and composite effects (fading + shadowing).Furthermore, the results of this paper are more complete analysis of our initial findings that were published in[11]. The major contributions are summarized as follows.

(i) The MGF based framework that characterizes the AUC metric requires only the k-th derivative of the MGF of the received SNR. It is simpler than the solution given in [5]. The key advantage of this framework is that the MGF of the end-to-end SNR is readily available for the common fading conditions or it can be computed easily.

(ii) The CDF based approach requires only the CDF of the end-to-end SNR. It involves only a single integral and the computation can be done simply by numerical approach. Moreover CDF of the common fading environments are readily available.

(iii) The developed frameworks can be used to analyze detectors' performance in diverse propagation channel modelsand can be easily extended to various *i.i.d/i.n.d.* diversity combining scenarios.

(iv) The solutions are capable of handling the half-odd integer value of u and m, which are in sharp contrast to most existing solutions.

(v) The solutions simplify the cases where many were either intractable or required complicated mathematical steps. Thus, the frameworks are simple andcomputationally efficient for analyzing detector's performance in myriad fading environments.

The rest of the paper is organized as follows. Section 2presentsthe system model for the energy detection. Section 3 includes proposed frameworks for the energy detector's performance analysis with the AUC metricfor no diversity scenarios. In Section 4, we extend the proposed frameworks for the diversity combining scenarios. Finally numerical results and concluding remarks are presented in Section 5and 6respectively.

2. SYSTEM MODEL

The observed waveform y(t) of the unknown deterministic signal s(t) at the receiver can be modeled by two hypotheses:H_0 and H_1;where hypothesis H_0 means that the signal is absent,whereas H_1 implies that the signal is present. The y(t) can be given as[12]

$$y(t) = \begin{cases} n(t) & : H_0 \\ hs(t) + n(t) & : H_1 \end{cases} \tag{1}$$

where,n(t) is the additive white Gaussian noise (AWGN), h is thecomplex channel gain between the primary signal transmitter and the detector. The energy detection is performed by comparing the measured energy of the observed signal in the observation time interval T with the energy threshold λ. The decision statistic Y can be represented as a Chi-square distribution χ^2_{2u} under H_0 and a noncentralChi-square distribution $\chi^2_{2u}(2\gamma)$ with $2u$ degree of freedom andnoncentrality parameter of 2γ under hypothesis H_1[9]. γ is the end-to-end instantaneous SNR and $u = TW$ is the time-bandwidth product (W is the filter bandwidth). Therefore, PDF of random variable Y under the two hypotheses can be written as

$$f_Y(y) = \begin{cases} \dfrac{y^{u-1}e^{-y/2}}{2^u\,\Gamma(u)}, & H_0 \\ \dfrac{1}{2}\left(\dfrac{y}{2u\gamma_i}\right)^{(u-1)/2} e^{-(y+2u\gamma)/2} I_{u-1}\left(\sqrt{2u\gamma y}\right), & H_1 \end{cases} \tag{2}$$

where, $\Gamma(\cdot)$ is the gamma function and $I_{u-1}(\cdot)$ is the modified Bessel function of the first kind. Thus, the probability of the false alarm $P_f = \mathrm{Pr}\,ob\{Y > \lambda \mid H_0\}$ and the probability of the detection $P_d = \mathrm{Pr}\,ob\{Y > \lambda \mid H_1\}$ can be derived respectively by integrating the PDF(for H_0 and H_1 Hypotheses) with limit λ to ∞ as[13]

$$P_f(\lambda) = \Gamma(u, \frac{\lambda}{2})/\Gamma(u) \tag{3}$$

$$P_d(\gamma, \lambda) = Q_u(\sqrt{2\gamma}, \sqrt{\lambda}) \tag{4}$$

where, $Q_u(.,.)$ is the generalized (u-thorder) Marcum Q-function and $\Gamma(\cdot,\cdot)$ is the upper incomplete Gamma function.

3. ENERGY DETECTION OVER MYRIAD FADING CHANNEL MODELS: AUC APPROACH

In this Section, we develop the proposed frameworks considering the AUC performance metric. We first derive the MGF based framework and then derive the CDF based simplified frameworks.

3. 1 MGF Based Framework

AUC varies from 0.5 to 1 as the energy detector threshold λ varies from ∞ to 0 [5].For a $P_d(\gamma, \lambda)$ vs. $P_f(\lambda)$ - ROC curve, instantaneous AUC can be expressed in term of instantaneous SNR γ as

$$A(\gamma) = \int_0^1 P_d(\gamma, \lambda)dP_f(\lambda) \tag{5}$$

Since $P_f(\lambda)$ and $P_d(\gamma, \lambda)$ are both functions of the threshold λ, the threshold averaging method[14] can be used in evaluating AUC. When $P_f(\lambda)$ varies from $0 \to 1$, λ can be seen to vary from $\infty \to 0$. Thus (5) can be re-written as

$$A(\gamma) = -\int_0^\infty P_d(\gamma, \lambda)\frac{\partial P_f(\lambda)}{\partial \lambda}d\lambda. \tag{6}$$

Taking the derivative of (3) and inserting into (6) we obtain

$$A(\gamma) = \frac{1}{2^u\,\Gamma(u)}\int_0^\infty \lambda^{u-1}e^{-\frac{\lambda}{2}}Q_u\left(\sqrt{2\gamma}, \sqrt{\lambda}\right)d\lambda \tag{7}$$

The detection probability (conditional probability) over the AWGN can be written using the canonical series representation of the generalized Marcum-Q function $Q_u(.,.)$ [15]as (8).

$$P_d(\gamma,\lambda) = Q_u\left(\sqrt{2\gamma},\sqrt{\lambda}\right) = \sum_{k=0}^{\infty} \frac{\gamma^k e^{-\gamma}}{k!} \frac{\Gamma\left(u+k,\frac{\lambda}{2}\right)}{\Gamma(u+k)} \tag{8}$$

An alternative canonical series representation of $Q_u(.,.)$can be derived by substituting $G(a,z) = \Gamma(a) - \Gamma(a,z)$ in (8), as

$$P_d(\gamma,\lambda) = Q_u\left(\sqrt{2\gamma},\sqrt{\lambda}\right) = 1 - \sum_{k=0}^{\infty} \frac{\gamma^k e^{-\gamma}}{k!} \frac{G\left(u+k,\frac{\lambda}{2}\right)}{\Gamma(u+k)} \tag{9}$$

where, $G(a,z)$ is the lower incomplete Gamma function. Itcan be noted here that the alternative Marcum Q-functionis capable of handling the half-odd positive integer $u$and m [15].Thus (8) or (9) also hold this property congenially. Later we show that both (8) and (9) are in desirable forms that facilitate the fading averaging problem in a unified manner using the MGF-derivative approach. Now substituting (8) or (9) into (7) we can obtain the unfaded AUCrespectively as

$$A(\gamma) = \frac{1}{2^u \Gamma(u)} \sum_{k=0}^{\infty} \frac{\gamma^k e^{-\gamma}}{k!\Gamma(u+k)} \frac{\Gamma(k+2u)_2 F_1(1,k+2u;1+u;0.5)}{u 2^{(u+k)}} \tag{10}$$

$$A(\gamma) = 1 - \frac{1}{\Gamma(u)} \sum_{k=0}^{\infty} \frac{\Gamma(k+2u)_2 F_1(1,k+2u;1+u+k;0.5)}{(u+k)\Gamma(k+u)k! \, 2^{(2u+k)}} \tag{11}$$

where, $_2F_1(.,.;.;.)$ is the confluent hypergeometric function. The average AUC $(\overline{A})$ over fading channels can be evaluated as

$$\overline{A} = \int_0^{\infty} A(\gamma) f_\gamma(\gamma) d\gamma \tag{12}$$

Substituting (10) into (12) yields average AUC as

$$\overline{A} = \frac{1}{u 2^u \Gamma(u)} \sum_{k=0}^{\infty} \frac{\Gamma(k+2u)_2 F_1(1,k+2u;1+u;0.5)}{\Gamma(k+u)k! 2^{(u+k)}} \int_0^{\infty} \gamma^k e^{-\gamma} f_\gamma(\gamma) d\gamma \tag{13}$$

where,

$$\int_0^{\infty} \gamma^k e^{-\gamma} f_\gamma(\gamma) d\gamma = (-1)^k \phi_\gamma^{(k)}(s)|_{s=1} \tag{14}$$

is the Laplace-transform identity. $\phi_\gamma^{(k)}(s)$is the$k$-th derivative of the MGF $\phi_\gamma(s)$ of the received SNR over the fading channel.Substituting (14) into (13), we obtain the average AUC over the generalized channels for the no-diversity as

$$\overline{A} = \frac{1}{u 2^u \Gamma(u)} \sum_{k=0}^{\infty} \frac{\Gamma(k+2u)_2 F_1(1,k+2u;1+u;0.5)}{\Gamma(k+u)k! 2^{(u+k)}} (-1)^k \phi_\gamma^{(k)}(s)|_{s=1} \tag{15}$$

Similarly, another generic expression for the average AUC can be derived using (11) as (16).

$$\overline{A} = 1 - \frac{1}{\Gamma(u)} \sum_{k=0}^{\infty} \frac{\Gamma(k+2u)\,_2F_1(1,k+2u;1+u+k;0.5)}{(u+k)\Gamma(k+u)k!2^{(2u+k)}} (-1)^k \phi_\gamma^{(k)}(s)|_{s=1} \qquad (16)$$

Since$k$-th derivative of the MGF of the common multipath fading channelsare available, the average AUC can be readily obtained from (15) or (16). Thus,(15) and (16) provides generalizedsolutions over the myriad fading environments. For convenience, the$k$-th orderderivativesof the MGF of the end-to-end SNR of common fading channels are listed in Table 1[16]. Similarly, (15) and (16) are also applicable to the channels with composite effects (fading with shadowing). To obtain the detection performance for channels with composite effects, we just need the higher order derivative of the MGF of the effective SNR of the channel with the composite effects.

Even though we have infinite series in (15) and (16), they converge with reasonable numbers of terms with truncation error less than .01%. Table 2 shows the number of terms required for obtaining a four figure accuracy for various channels. The Table also shows that the series in (16) converges much faster than the series in (15). Herath et al. also showed the convergence for the similar Marcum-Q function in [17] with reasonable number of terms.

It is now worth mentioning that the MGF based solution also simplifies the ROC analysis. We can easily obtain the average detection probability $\overline{P}_d(\gamma,\lambda)$ by averaging the $P_d(\gamma,\lambda)$ ((8) or (9)) over the PDF of channel SNR and express in terms of the higher order derivatives of the MGF of the end-to-end SNR as

$$\overline{P}_d(\gamma,\lambda) = \sum_{k=0}^{\infty} \frac{(-1)^k}{k!} \frac{\Gamma\left(u+k,\frac{\lambda}{2}\right)}{\Gamma(u+k)} \phi_\gamma^{(k)}(s)\Big|_{s=1} \qquad (17)$$

$$\overline{P}_d(\gamma,\lambda) = 1 - \sum_{k=0}^{\infty} \frac{(-1)^k}{k!} \frac{G\left(u+k,\frac{\lambda}{2}\right)}{\Gamma(u+k)} \phi_\gamma^{(k)}(s)\Big|_{s=1} \qquad (18)$$

Thus, using Table-I, we can also easily obtain $\overline{P}_d(\gamma,\lambda)$ for various channel models.

3.2 CDF Based Framework

We were motivated to develop a CDF based framework to handle specific diversity combining scenarios. For some diversity combining cases CDF based approach has advantages over other frameworks;selection combining (SC) is such a scheme. The average AUC, based on the CDF approach can be given as (See the detail derivation in Appendix-A).

$$\overline{A} = 1 - \frac{\Gamma(2u)}{2^{2u}\Gamma(u)\Gamma(u+1)} \int_0^{\infty} F_\gamma(\gamma)e^{-\gamma} {}_1F_1(2u,u+1;\frac{\gamma}{2})d\gamma \qquad (19)$$

Eqn. (19) can be solved numerically for fading channels using the CDF listed in [18, pp. 420] easily by common mathematical software packages (e.g., MATLAB, Mathematica).

Table 1.MGFof SNRand its k-th Derivative for Several Common Stochastic Channel Models

Channel Model	MGF of SNR $\phi_\gamma(s)=\int\limits_0^\infty f(\gamma)e^{-s\gamma}d\gamma$
	k^{th} Derivative of the MGF $\phi_\gamma^{(k)}(s)$ with L-Branch (i.i.d.)(L=1: no-diversity)
Rayleigh	$(1+s\Omega)^{-1}$
	$(-\Omega)^k\Gamma(L+k)/\left((1+s\Omega)^{L+k}\Gamma(L)\right)$
Nakagami –q $b=\dfrac{1-q^2}{1+q^2}$	$\left(1+2s\Omega+4s^2\Omega^2q^2/(1+q^2)^2\right)^{-1/2}$
	$\dfrac{k!}{[\Gamma(L/2)]^2\sqrt{[1+s\Omega(1-b)]^{L/2}}[1+s\Omega(1+b)]^{L/2}}\left[\dfrac{-\Omega(1+b)}{1+s\Omega(1+b)}\right]^k$ $\times\sum\limits_{w=0}^k\dfrac{\Gamma(L/2+w)\Gamma(L/2+k-w)}{[w!(k-w)!]^2}\left[\dfrac{(1+b)[1+s\Omega(1+b)]}{(1+b)[1+s\Omega(1-b)]}\right]^w$
Nakagami–n (Rice: K=n²)	$\dfrac{1+K}{1+K+s\Omega}\exp\left(\dfrac{-Ks\Omega}{1+K+s\Omega}\right)$
	$\dfrac{(-\Omega)^k k!(L+k-1)!(1+K)^L}{(1+K+s\Omega)^{L+k}}\exp\left(\dfrac{-sKL\Omega}{1+K+s\Omega}\right)\sum\limits_{i=0}^k\dfrac{1}{i!(L+i-1)!(k-i)!}\left(\dfrac{KL(1+K)}{1+K+s\Omega}\right)^i$
Nakagami-m	$\left(1+\dfrac{s\Omega}{m}\right)^{-m}$
	$\dfrac{(-\Omega)^k m^{mL}\Gamma(mL+k)}{(m+s\Omega)^{mL+k}\Gamma(mL)}$
K_GDistribution	$\left(\dfrac{\Theta}{s}\right)^{\beta/2}\exp\left(\dfrac{\Theta}{2s}\right)W_{-\beta/2,\alpha/2}\left(\dfrac{\Theta}{s}\right)$ $\sum\limits_{n=1}^k\sum\limits_{p=0}^n\sum\limits_{j=0}^n\binom{n}{j}\dfrac{1}{n!}\dfrac{(-1)^{p+k+j}(n)_{\bar p}}{p!}(\Theta^{a+j}/s^{a+j+k})(n-p)_k(a)_j(a)_{\overline{n-j}}\Psi(a+j;1+\alpha+j;y);y),\quad k>0$ $\phi_\gamma^{(0)}(s)=\phi_\gamma(s),\quad k=0$ $\Theta=km/\Omega,\ \alpha=k-m,\ \beta=k+m-1,\ a=(\alpha+\beta+1)/2,$ $\Psi(\cdot;,;.)=\text{Tricomi confluent hyper. func. }(a)_n=a(a+1)(a+2).....(a+n-1)\ ;$ $(a)_{\bar n}=a(a-1)(a-2)....(a-n+1)\ W_{\lambda,\mu}(.)=\text{Whittaker function.}$
Weibull	$\left(\dfrac{c}{\beta}\right)(2\pi)^{\frac{1-c}{2}}c_i^{(c-1/2)}s^{-c}G_{1,c}^{c,1}\left(\beta\left(\dfrac{s}{c}\right)^c\left|\begin{matrix}1\\1,1+\dfrac{1}{c},...1+\dfrac{c-1}{c}\end{matrix}\right.\right);\ 0\prec c$
	$\left(\dfrac{\Gamma(1+2/c)}{\Omega}\right)^{c/2}(2\pi)^{(1-c/2)/2}\left(\dfrac{c}{2}\right)^{k+\frac{c}{2}+\frac{1}{2}}s^{-(k+\frac{c}{2}+1)}G_{1,c/2}^{c/2,1}\left(\left(\dfrac{2s\Gamma(1+2/c)}{c\Omega}\right)^{-c/2}\left|\begin{matrix}1\\1+\dfrac{2k}{c},1+\dfrac{2(k+1)}{c},...2+\dfrac{2(k-1)}{c}\end{matrix}\right.\right)$
G- Dist	$\phi_\gamma(s)=m\sum\limits_{r=0}^{m+k-1}\binom{m+k-1}{r}\dfrac{(-1)^{r+1}s^{-k}}{(r+1-k)!}\sum\limits_{p=0}^r\binom{r}{p}\left(2\sqrt{\dfrac{\alpha s}{\beta}}\right)^{r-p+1}\Gamma(r+p+1)H_{-(r+p+1)}(\dfrac{b}{2}\sqrt{\dfrac{b}{s}}+\sqrt{\dfrac{\alpha s}{\beta}})$
	$m\sum\limits_{r=0}^{m+k-1}\binom{m+k-1}{r}\dfrac{(-1)^{r+1}s^{-k}}{(r+1-k)!}\sum\limits_{p=0}^r\binom{r}{p}\left(2\sqrt{\dfrac{\alpha s}{\beta}}\right)^{r-p+1}\Gamma(r+p+1)H_{-(r+p+1)}(\dfrac{b}{2}\sqrt{\dfrac{b}{s}}+\sqrt{\dfrac{\alpha s}{\beta}}),\quad k\succ 0$ $\phi_\gamma^{(0)}(s)=\phi_\gamma(s),\quad k=0\ ;\ H_\nu(x)=\text{Hermite function}$

*L-BRANCH for MRC/SLC

Table 2.Number of Terms Required to Obtain Four Figure Accuracy (N)

Channel	SNR= -5 dB u = 2	SNR= -5 dB u = 4	SNR= 5 dB u = 2	SNR= 5 dB u = 4	SNR= 10 dB u = 2	SNR = 10 dB u = 4
$\left| E_{Nak} \right|$	m =4, N_1 = 6 m =4, N_2 = 4	m = 4, N_1 = 6 m =4, N_2 = 5	m = 4, N_1 = 23 m = 4, N_2 = 9	m =4, N_1 = 19 m =4, N_2 = 11	m =4, N_1 = 49 m =4, N_2 = 13	m =4, N_1 = 47 m =4, N_2 = 16
$\left| E_{Nak-q} \right|$	q = 1, N_1 = 8 q = 1, N_2 = 5	q = 1, N_1 = 8 q = 1, N_2 = 5	q = 1, N_1 = 40 q = 1, N_2 = 10	q = 1, N_1 = 38 q = 1, N_2 = 12	q = 1, N_1 = 127 q = 1, N_2 = 10	q = 1, N_1 = 119 q = 1, N_2 = 13
$\left| E_{Rice} \right|$	K = 4, N_1 = 6 K = 4, N_2 = 4	K = 4, N_1 = 6 K = 4, N_2 = 5	K = 4, N_1 = 21 K = 4, N_2 = 11	K = 4, N_1 = 21 K = 4, N_2 = 12	K = 4, N_1 = 52 K = 4, N_2 = 13	K = 4, N_1 = 50 K = 4, N_2 = 14
$\left| E_{Nak,2(MRC)} \right|$	m = 4, k=1, N_1= 7 m = 4, k=1, N_2= 5	m = 4, k=1, N_1= 7 m = 4, k=2, N_2= 6	m = 4, k=1, N_1= 26 m = 4, k=2, N_2= 12	m = 4, k=1, N_1= 26 m = 4, k=2, N_2= 12	m = 4, k=1, N_1= 66 m = 4, k=2, N_2= 11	m = 4, k=1, N_1= 65 m = 4, k=2, N_2= 14
$\left| E_{Rice,2(MRC)} \right|$	K = 2, N_1 = 7 K = 2, N_2 = 5	K = 2, N_1 = 7 K = 2, N_2 = 7	K = 2, N_1 = 32 K = 2, N_2 = 11	K = 2, N_1 = 31 K = 2, N_2 = 13	K = 2 , N_1 = 88 K = 2, N_2 = 13	K = 2, N_1 = 87 K = 2, N_2 = 17

* N_1 represents Eq. (15); N2represents Eq. (16)

4. DIVERSITY BASED ENERGY DETECTOR

Energy detector's performance is severely limited by harsh propagation environments and becomes unreliable at low signal-to-noise ratio (SNR). Diversity combining is proposed in many works to overcome this limitation.In this Section, we have studied the diversity combining based energy detection using the frameworks developed in Section 3. Diversity based receiver receives redundantly the same information-bearing signal over multipath channels and combines multiple replicas in order to increase the overall received SNR. The combining can be done in many ways, which varies in complexity and overall performance. Common diversity reception can be done via space, frequency, time, and delay diversity[18].Diversity combining that takes place at RF is called pre-detection combining, while diversity combining that takes place at the baseband is called post-detection combining. In many cases, there is no difference in performance, at least in an ideal sense.To show the versatility of our frameworks and the efficacy of the diversity combining, we illustrate the energy detector's performance for MRC, SLC, SC,switch and stay combining (SSC) schemes for L independent branchesover the Nakagami-m and or Rice channels for flat fading components and over K_G distribution for the composite effects in the following sub-sections.

4.1. Maximum Ratio Combining (MRC)

The output SNR, γ_{MRC} of the MRC combiner is the sum of all the SNRs on all branches, i.e.,

$\gamma_{MRC} = \sum_{l=1}^{L} \gamma_l$; where L is the number of diversity branches[18]. The statistics of the decision variable of a MRC follows χ_{2u}^2 distribution under H_0 and $\chi_{2u}^2(\gamma_{MRC})$ under H_1 Hypothesis[9]. Therefore, the $P_f(\lambda)$ and the $P_d(\gamma,\lambda)$ at the MRC output for AWGN channels can be evaluated by (3) and (4) respectively. Thus, using the TABLE 1 for the $i.i.d.L$-branch, average AUC over various channels can be readily derived from (15) or (16). For instance, average AUC for $i.i.d.L$ branches over Nakagami-m and Rice channels models can be given in (20) and (21) respectively.

$$\overline{A}(\gamma)_{Nak}^{MRC} = \frac{1}{u2^u\,\Gamma(u)} \sum_{k=0}^{\infty} \frac{\Gamma(k+2u)\,_2F_1(1,k+2u;1+u;0.5)}{\Gamma(k+u)k!2^{(u+k)}} \frac{\Omega^k m^{mL}\Gamma(mL+k)}{(m+\Omega)^{mL+k}\,\Gamma(mL)} \qquad (20)$$

$$\overline{A}(\gamma)_{Ric}^{MRC} = \frac{1}{u2^u\Gamma(u)}\sum_{k=0}^{\infty}\frac{\Gamma(k+2u)\,_2F_1(1,k+2u;1+u;0.5)}{\Gamma(k+u)k!2^{(u+k)}}\frac{(\Omega)^k\,k!(L+k-1)!(1+K)^L}{(1+K+\Omega)^{L+k}}$$

$$\times\exp\left(\frac{-KL\Omega}{1+K+\Omega}\right)\sum_{i=0}^{k}\frac{1}{i!(L+i-1)!(k-i)!}\left[\frac{KL(1+K)}{1+K+\Omega}\right]^i \qquad (21)$$

For general cases of *i.n.d.* branches, we can use the MGF of the effective SNR at the combiner as $\phi(s)_{MRC} = \prod_{i=1}^{L}\phi(s)_i$, where $\phi(s)_i$ is the MGF of the SNR of the *i*-th branch. Using the Leibniz differentiation rule [19, eq. (0.42)], we can obtain the *k*-th derivative of $\phi(s)_{MRC}$ as

$$\phi(s)_{MRC}^{(k)} = \sum_{n_1=0}^{k}\sum_{n_2=0}^{n_1}\cdots\cdots\sum_{n_{L-1}=0}^{n_{L-2}}\binom{k}{n_1}\binom{n_1}{n_2}\cdots\binom{n_{L-2}}{n_{L-1}}\phi(s)_1^{(k-n_1)}\phi(s)_2^{(k-n_2)}(s)\ldots\ldots\ldots\phi(s)_L^{(n_{L-1})} \qquad (22)$$

Thus, using (22) in (15) or (16), average AUC can be obtained for MRC over *i.n.d.* branches for any fading channel.

4.2. Square-Law Combining (SLC)

Since the output decision is combined after the sampling, the decision variable is the sum of *Li.i.d.* The statistics of the decision variable of a SLC follows χ_{2u}^2 distribution under H_0 and $\chi_{2Lu}^2(\varepsilon_{SLC})$ under H_1 Hypothesis[9];where the non-centrality $\varepsilon_{SLC} = \sum_{l=1}^{L}\varepsilon_l = \sum_{l=1}^{L}2\gamma_l = 2\gamma_{SLC}$;hence the $P_f(\lambda)$ and $P_d(\gamma,\lambda)$ can be expressed by (3) and (4) respectively by replacing u with Lu. Thus, for the SLC scheme, average AUC can be obtained over myriad fading environments using the same derived expressions (15) or (16) just by replacing u with Lu.Similar to the *i.n.d.* MRC, we can also obtain average AUC for the *i.n.d.*SLCas $\phi(s)_{SLC} = \phi(s)_{MRC} = \prod_{i=1}^{L}\phi(s)_i$.

4.3. Selection Combining (SC)

With SC, the path output from the combiner has an SNR equal to the maximum SNR of all the branches given by $\gamma_{sc} = \max(\gamma_1,\gamma_1,...,\gamma_L)$. Because of the single branch considered at any given period the decision output is the *i.i.d.* variable and follows the χ_{2u}^2 forH_0 and $\chi_{2u}^2(\varepsilon_j)$ forH_1.Since SC selects single branch at any given period, we have treated this scheme based on the developed CDF approach. For L branch diversity, the CDF is given as[18]

$$F_{\gamma_{sc}} = P(\max[\gamma_1,\gamma_2,....\gamma_L]<\gamma) = \prod_{l=1}^{L}P(\gamma_l<\gamma) = \prod_{l=1}^{L}F_{\gamma_l}(\gamma) \qquad (23)$$

where, $F_{\gamma_l}(\gamma)$ is the CDF of the *l*-th branch. Thus, the average AUC over the fading channels for the SC scenario can be given readily using (19) as

$$\overline{A}^{SC} = 1 - \frac{\Gamma(2u)}{2^{2u}\Gamma(u)\Gamma(u+1)}\int_{0}^{\infty}F_{\gamma_{sc}}(\gamma)e^{-\gamma}\,_1F_1(2u,u+1;\frac{\gamma}{2})d\gamma \qquad (24)$$

As an illustration, average AUC can be given using the CDF [18] of link SNR over*i.n.d.* Nakagami-*m* channels for the SC scheme from (24) as

$$\overline{A}_{Nak}^{SC} = 1 - \frac{\Gamma(2u)}{2^{2u}\Gamma(u)\Gamma(u+1)} \int_0^\infty \prod_{l=1}^L \left[1 - \Gamma\left(m_l, \frac{m_l}{\Omega_l}\gamma\right)/\Gamma(m_l)\right] e^{-\gamma} {}_1F_1(2u,u+1;\frac{\gamma}{2})d\gamma \quad (25)$$

For *i.i.d.* case, (25) becomes

$$\overline{A}_{Nak}^{SC} = 1 - \frac{\Gamma(2u)}{2^{2u}\Gamma(u)\Gamma(u+1)} \int_0^\infty \left[1 - \Gamma\left(m, \frac{m}{\Omega}\gamma\right)/\Gamma(m)\right]^L e^{-\gamma} {}_1F_1(2u,u+1;\frac{\gamma}{2})d\gamma. \quad (26)$$

Similarly the average AUC over the *i.i.d.* Rice channels can be given as

$$\overline{A}_{Ric}^{SC} = 1 - \frac{\Gamma(2u)}{2^{2u}\Gamma(u)\Gamma(u+1)} \int_0^\infty \left[1 - Q_1\left(\sqrt{2K}, \sqrt{\frac{2(1+K)\gamma}{\Omega}}\right)\right]^L e^{-\gamma} {}_1F_1(2u,u+1;\frac{\gamma}{2})d\gamma \quad (27)$$

where, $Q_1(.,.)$ is the first order Marcum Q function. Although in[5], closed form expression is developed for SC scheme, it requires critical mathematical calculationswhile our CDF based approach is simple and can be solved numerically using any math package easily.Hencethis framework is more attractive compared to the one given in [5]. Moreover, the developed expression is applicable to all common fading channels. This method can also be usedsuitably for characterizing the ROC curve of average detection probability.

4.4. Composite Channels

Wireless channel is a complicated phenomenon and the communication is affected by not only various multipath fading but also by shadowing effects. Shadowing process is typically modelled as a lognormal distribution [20]. Practical wireless channels can be modelled as multipath fading superimposed on lognormal shadowing but due to the difficulty of analyzing, shadowing effect is commonly neglected in the literature. But in recent time, Shankar[21] showed that the generalized-$K(K_G)$ distribution can be used to model fading and shadowing phenomena effect in mobile communication channels and this distribution makes the mathematical performance analysis much simpler to handle as compared to Lognormal-based models such as Nakagami-m or the Rayleigh-Lognormal (R-L) model.To take this advantage, we considered the $K_G$distribution to include the composite effects in the performance analysis.

Similar to the multipath fading scenarioswe can also derive the average AUC over the composite channels (*i.i.d.* or *i.n.d.* cases) for various combining schemes. For example, the SLCover *i.n.d.*K_G distribution for composite effects can be given as

$$\overline{A}_{comp}^{SLC} = 1 - \frac{1}{\Gamma(uL)} \sum_{k=0}^\infty \frac{\Gamma(k+2uL){}_2F_1(1,k+2uL;1+uL+k;0.5)}{(uL+k)\Gamma(k+uL)k!2^{(2uL+k)}}(-1)^k$$

$$\times \sum_{n_1=0}^k \sum_{n_2=0}^{n_1} ...... \sum_{n_{L-1}=0}^{n_{L-2}} \binom{k}{n_1}\binom{n_1}{n_2}\binom{n_2}{n_3}....\binom{n_{L-2}}{n_{L-1}} \phi_1^{(k-n_1)}(s)|_{s=1}\ \phi_2^{(n_1-n_2)}(s)|_{s=1}......\phi_L^{(n_{L-1})}(s)|_{s=1}$$

$$(28)$$

where, $\phi_i^{(k)}(s)$ is the k-th derivative of the MGF of the i-th channel SNR with composite effects (see Table-1,K_G distribution).For SC scheme, we follow the derived CDF approach (19). The CDF $F_\gamma(\gamma)$ can be given as [22]

$$F_\gamma(\gamma) = \pi \csc(\pi\alpha) \left[\frac{(\Theta\gamma)^m \, _1F_2(m;1-\alpha,1+m;\Theta\gamma)}{\Gamma(k)\Gamma(1-\alpha)\Gamma(1+m)} - \frac{(\Theta\gamma)^k \, _1F_2(k;1+\alpha,1+k;\Theta\gamma)}{\Gamma(m)\Gamma(1+\alpha)\Gamma(1+k)} \right] \quad (29)$$

where, pFq ($\cdot$) is the generalized hypergeometric function[19, eq. (9.14/1)], and p, q are integers. Thus, inserting (29) in to (24), average AUC for SC scheme can be derived over the K_G distribution easily by numerical methods.

5. NUMERICAL RESULTS

We have used Mathematica and MATLAB software to obtain various numerical results of the average AUC in myriad channel environments.We completed analyses in two parts; the first part includes the no-diversity and the second part includes the diversity combining scenarios for *i.i.d./i.n.d.*channels. But first we want to give a glimpse of AUC's appealing approachin contrast to ROC curve. We show how the AUC works as the single figure of merit and useful it could be specially when the ROC curves overlaps.Figure 1 shows the overlapping complementary ROC curves for various detectors. It is difficult to picture visually which detector gives better performance in the case of overlapping scenario. Using the average AUC expressions proposed in this article, we obtain distinct average AUC curves against the SNR, which makes the comparison much easier (Figure 2). The computed average AUC value at mean 10 dB for the curves (top to bottom) in Figure2 are 0.9966, 0.9898, 0.9821, 0.9663, 0.924, 0.9158. This implies that the average AUC can be a single figure of merit that characterizes the performance of an energy detector over the range of received signal SNR.

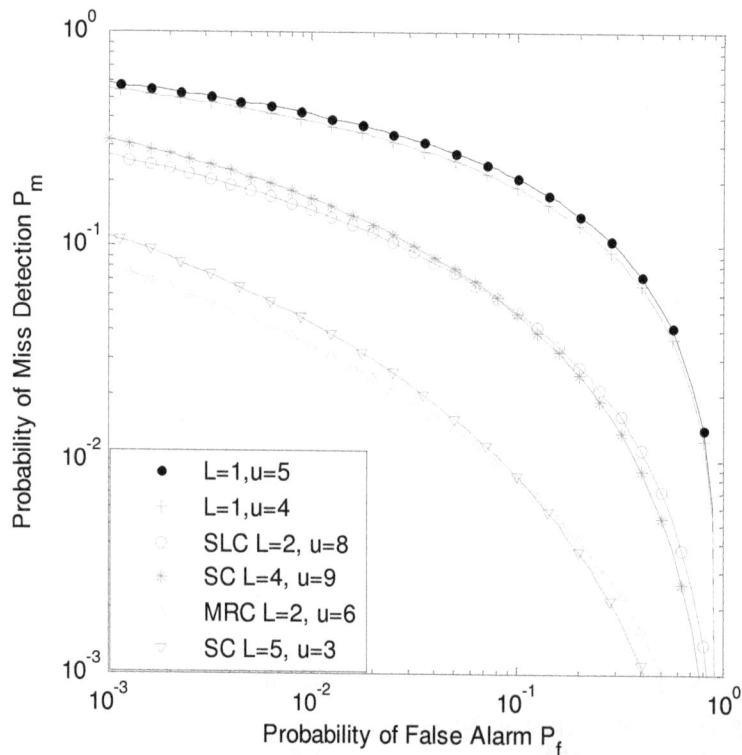

Figure 1. Complementary ROC curves of different detection configuration in Nakagami-*m* channel, *m=2, mean SNR = 10 dB*

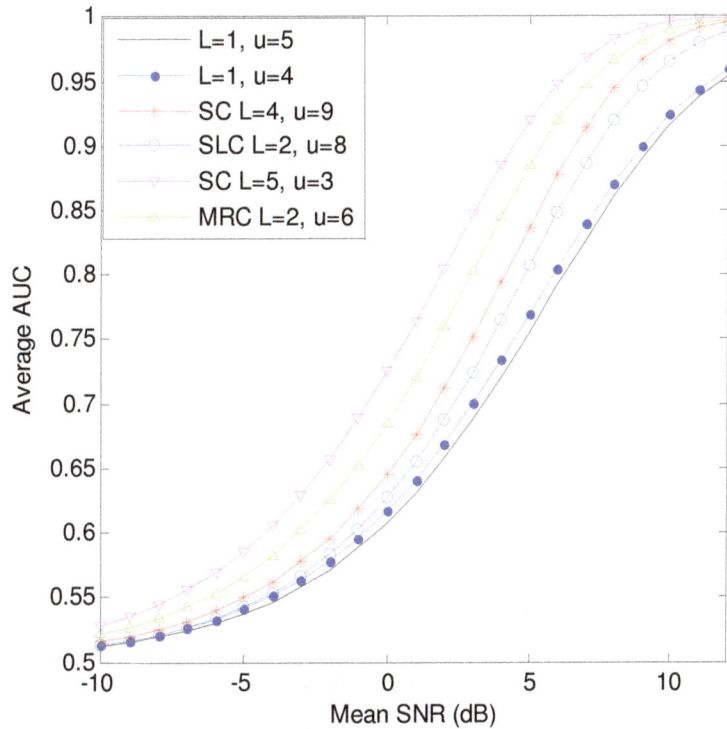

Figure 2. Average AUC vs. mean SNR for various energy detectors in Nak-*m* channels (m=2)

5.1. No-Diversity Scenarios

In Figure 3, we present the strength of the MGF based framework by showing the performance curves in various fading channel models. The detection performance in[5] with AUC metric is analyzed only over Nakagami-*m* fading channel and the expression is not easy to extend for generalization whereas our derived frameworks are compact and can be easily utilized for various *i.i.d./i.n.d.* fading channels.For illustrations we have considered the Nakagami-*m*,Rice, Nakagami-q and $K_G$distribution. Each AUC curve in Figure3shows that the performance becomes 0.5 if the detector's performance is no better than flipping a coin and increases to 1 as the performance improves. Thus, it is more meaningful to describe practical detector's performance with the AUC metric in contrast to ROC metric. Although not reported here, we noticed that at high SNR, the performance is better but for a fixed SNR, the higher sampling (*u*)numbers do not increase the detection performance since the false alarm also increases.We also observed though not shown that the impact of Rice K factor on the performance is insignificantfor curves below 0 dB and it affects slightly on curves over 5 dB. When shadowing is present with the multipath fading effects, the detection performance is not as good as the case when only multipath fading is present. This can be seen by comparing the curve for the composite effects (K_G distribution) and the Nakagami-m channel only.   When *m*=1, $K_G$distribution becomes K distribution. We also noticed that the detection performance is better over $K_G$distribution in comparison to only$K$ distribution (not shown here).Figure 3also includes the curves with the half-odd integer *u* and *m* values, which are in sharp contrast for most existing results.

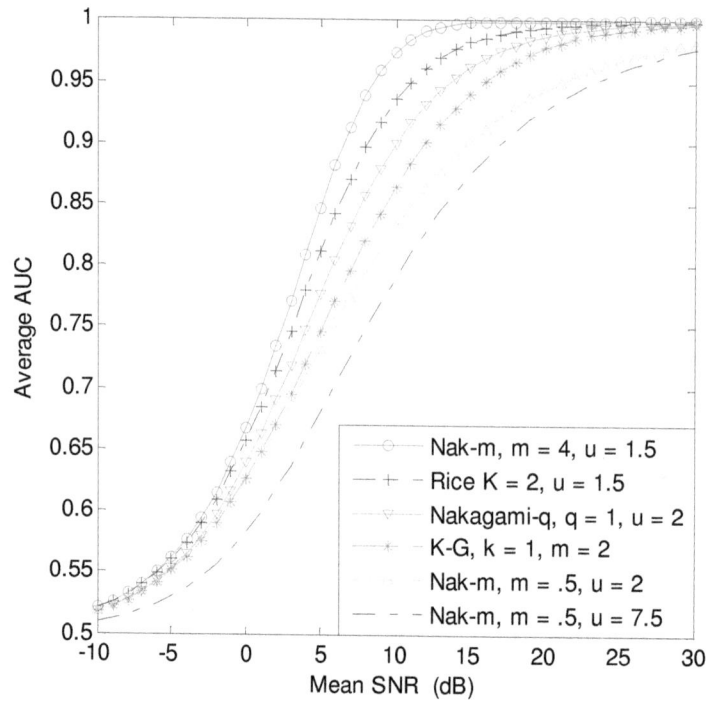

Figure 3. Average AUC vs. mean SNR for no-diversity

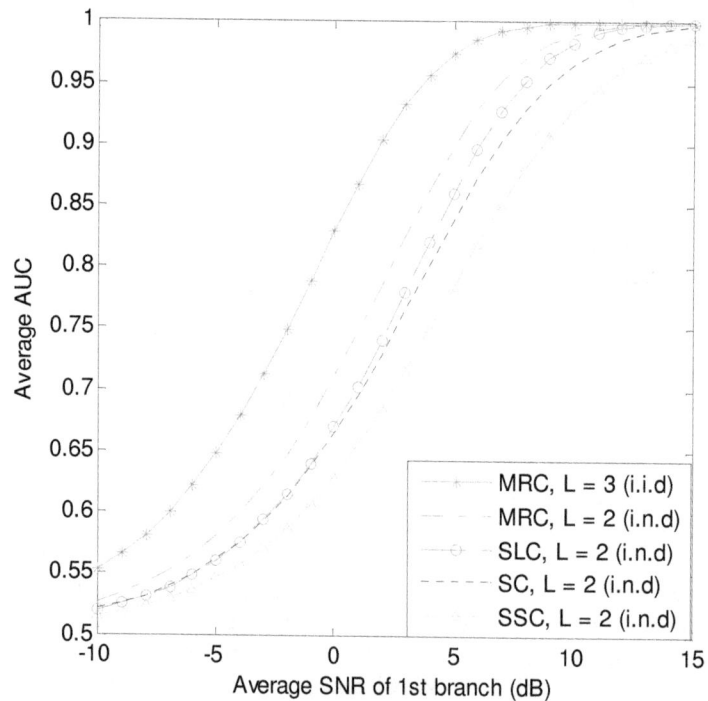

Figure 4. Average AUC vs. mean SNR with diversity combining over Nakagami-m channels; $m_1 = 3$, $m_2 = 2$ $u = 2$, $\bar{\gamma}_1 = 2\bar{\gamma}_2$, where $\bar{\gamma}_i$ is the end-to-end mean SNR for i-th branch.

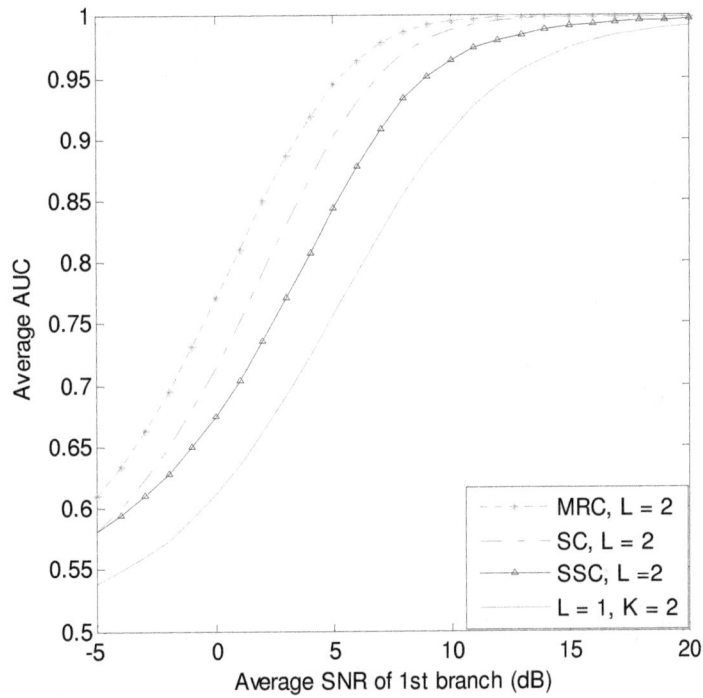

Figure 5. AUC vs. mean SNR with dual diversity combining over *i.n.d.* Rice fading
$(K_1 = 2, K_2 = 4, u = 4.5, \bar{\gamma}_2 = 2\bar{\gamma}_1)$.

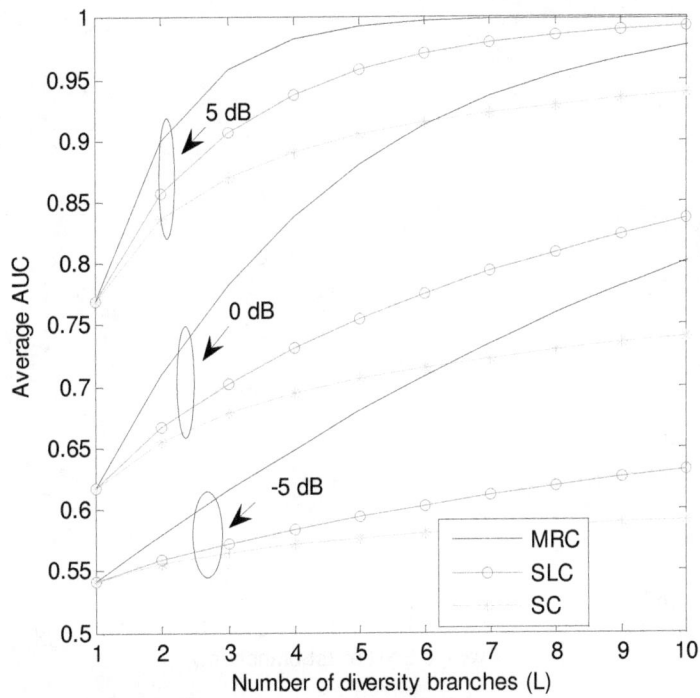

Figure 6. Average AUC vs. diversity branch L over *i.i.d.* Nak-m channel $(u = 4, m = 2)$

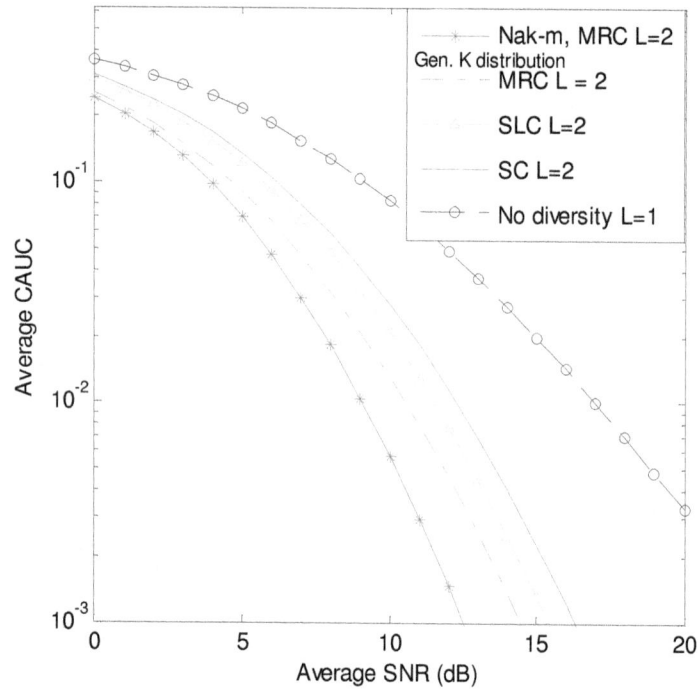

Figure 7.   Average Complementary AUC (CAUC) for dual diversity reception over K_G distribution ($u = 2, k = 3, m = 2$)

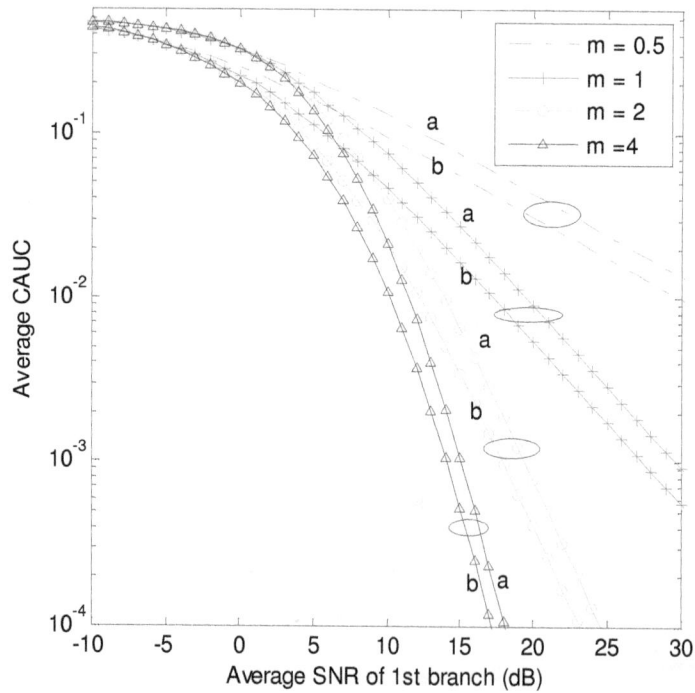

Figure 8. Comparison of (a) *i.i.d.* ($\bar{\gamma}_2 = \bar{\gamma}_1$) and (b) *i.n.d.* cases ($\bar{\gamma}_2 = 5\bar{\gamma}_1$) for SSC scheme over Nakagami-*m* channels

5.2. Diversity Scenarios

The developed frameworks can be applied easily for common receiver diversity combining scheme over myriad fading channels. For illustrations, we have presented results for MRC, SLC, SC and SSC schemes over Nakagami-m, Rice, and K_G distribution for *i.i.d./i.n.d.* cases. Mostly dual diversity combining is considered for simplicity. We used (22) either with (15) or (16) for *i.n.d.*MRC and SLC for demonstrating detection performance over various fading channels. For *SSC*, we used generalized framework (15) or (16) with the optimized threshold value[18]. For the *SC* scheme we used the CDF based solution given in (24).

In Figure 4, we have presented detection performance for various receiver diversity combining schemes over the *i.i.d.* and *i.n.d.* Nakagami-m fading channel model while in Figure5, we have produced results for various diversity combining over *i.n.d.* Rice fading channels. It can be noted now that our frameworks are straight forward to handle the *i.n.d.* cases, which were until now either intractable or intricate to solve because of the cumbersome frameworks. Figure 6 shows the impact of number of branches on the detection performance over the *i.i.d.* Nakagami-m channels. In Figure 6, it is observed that when the number of branch increases, the detection performance also increases but for higher SNR (> 5 dB), more than 5 branches do not make any significant difference in detection performance. In all cases (Figure4- Figure6),MRC shows the upper bounded performancebecause of the higher end-to-end SNR associated with the *MRC*but in reality MRC combiner is impracticalfor simple energy detector since it is a coherent combining method; instead SLC scheme is suitable for practical implementation. In Figure7, the dual diversity reception is illustrated for various combining schemes over the *i.i.d.*$K_G$distribution for the composite effects.A comparison is also made in this plot for the MRC scheme between theNakagami-m and K_G distribution. It shows that when the shadowing is present, detection performance is poor compared to one without the shadowing effect. Finally in Figure8, comparisonsare made by the complementary AUC curves between *i.i.d.* and *i.n.d.*dual diversity SSC scheme in Nakagami-m fading channels with different m values. It shows that as the m increases, the detection performance also improves as expected and the performance for*i.n.d.* shows betterfor cases when $\bar{\gamma}_2 = n\,\bar{\gamma}_1$ (where $\bar{\gamma}_i$ is the end-to-end mean SNR for *i*-th branch).

In all cases, diversity combining shows better detection probability vs. no-diversity cases even at the low SNR. The composite effects are usually neglected in the research communitydue to its mathematical complexity but our frameworks are capable of handling the analytical difficulty.Our frameworks establish that they can be used with any fading environments with even half-odd integer u or m values (fading index of Nakagami-m channels). Thus, we have demonstrated the usefulness of the proposed frameworks and also emphasized that the AUC performance metric is a more desirable metric in contrast to ROC for describing the energy detection performance measurement.

6. CONCLUSIONS

We have developed two analytical approaches to characterize the AUC performance metric of energy detectors over myriad fading channel models. We then study the usefulness of the frameworks for the diversity combining techniques in *i.i.d./i.n.d.* wireless channels subject to multipath fading and composite effects. The results indicated that our solutions are capable ofhandling myriad fading statistics for single channels as well as for various diversity combining techniques. The results also demonstrated that at low SNR the diversity combining techniques improve the signal detection probability without any rigorous mathematics. The results could be readily used in deciding the diversity order and/or the energy threshold value required to achieve a specified false alarm rate for different operating scenarios of a cognitive radio system.

We emphasized AUC metric to describe the energy detector's performance more meaningfully. Although AUC measurement provides simple and meaningful system performance, it is addressed only recently in the communication areas. Thus, our work adds significant values to the communication fields.The significant contribution of this work is the simplicity of our approachin analysing the energy detector's performance by involving either single fast converging series in conjunction with higher order derivative of the MGF of the channel SNR or by a CDF based single integral formula.The frameworksdo not require vigorous mathematical calculations and allow obtaining new results for several other cases for which simple solutions were not available until now. Moreover, our approach can be used with half-odd positive integer u and m values. To our best knowledge, use of half-odd positive integer u or half-odd positive integer $m$alone in the detection performance is not available in the literature besides our own works. Our frameworks can be easily extended for the relay based cooperative spectrum sensing and the results could readily be used to choose right parameters for energy detectors in cooperative and cognitive radio systems. Our work, thus has contributed significantly in cooperative,and cognitive communications areas.

APPENDIX-A

The average probability of detection can be given as

$$\overline{P_d} = \int_0^\infty Q_u\left(\sqrt{2\gamma}, \sqrt{\lambda}\right) f_\gamma(\gamma) d\gamma \tag{A-1}$$

where, $f_\gamma(\gamma)$ is the PDF of the end-to-end link SNR. Using the integration by part and the identity[24, eq. (11)], it is straight forward to show that

$$\overline{P_d} = 1 - \int_0^\infty \left(\frac{\lambda}{2\gamma}\right)^{u/2} e^{-\left(\gamma+\frac{\lambda}{2}\right)} I_u\left(\sqrt{2\gamma\lambda}\right) F_\gamma(\gamma) d\gamma \tag{A-2}$$

Where, $F_\gamma(\gamma)$ is the CDF of the link SNR and $I_u(\cdot)$ is the u-th order modified Bessel function of the first kind. Using (6) in (12), we can write:

$$\overline{A} = -\int_0^\infty \overline{P_d} \frac{dP_f(\lambda)}{d\lambda} d\lambda \tag{A-3}$$

Substituting (3) and (A-2) in to (A-3), we obtain

$$\overline{A} = -\int_0^\infty \frac{\partial P_f}{\partial \lambda} d\lambda - \frac{1}{2^u \Gamma(u)} \int_0^\infty F_\gamma(\gamma) e^{-\gamma} (2\gamma)^{-u/2} I(\lambda) d\gamma \tag{A-4}$$

where,
$$I(\lambda) = \int_0^\infty \lambda^{\frac{3}{2}u-1} e^{-\lambda} I_u\left(\sqrt{2\gamma\lambda}\right) d\lambda \tag{A-5}$$

Substituting $u = 2n$ and using [19, eq. (6.643-2)], (A-5) can be given in terms of u as (A-6).

$$I(\lambda) = \frac{\Gamma(2u)}{\Gamma(u+1)}\left(\frac{\gamma}{2}\right)^{-1/2} e^{\gamma/4} M_{\left(-\frac{3}{2}u+\frac{1}{2}\right),\frac{u}{2}}\left(\frac{\gamma}{2}\right)$$ (A-6)

where $M_{k,m}(z)$ is the Whittaker function and with the identity [18, eq. (9.220-2)], (A-6) can be simplified as

$$I(\lambda) = \frac{\Gamma(2u)}{\Gamma(u+1)}\left(\frac{\gamma}{2}\right)^{u/2} {}_1F_1(2u, u+1; \frac{\gamma}{2})$$ (A-7)

where, $_1F_1($ a, b ; c $)$ is the confluent hypergeometric function. Substituting (A-7) in (A-4), we obtain the average AUC over the fading channels as (19).

REFERENCES

[1]   S. Haykin, D. Thomson, and J. Reed, "Spectrum sensing for cognitive radio," *Proc. IEEE*, vol. 97, no. 5, pp. 849-877, May 2009.

[2]   K. Cheng and R. Prasad., *Spectrum sensing in cognitive radio networks*, 1st ed. New York, USA: John Wiley & Sons, 2009, pp. 184-229.

[3]   T. Fawcett, "An introduction to ROC analysis," *Pattern Recognition Letter*, vol. 27, no. 8, pp. 861-874, June 2006.

[4]   J. Hanley and B. Mcneil, "The meaning and use of the area under a receiver operating characteristics (ROC) curve," *Radiology*, vol. 143, no. 1, pp. 29-36, Apr. 1982.

[5]   S. Atapattu, C. Tellambura, and H. Jiang, "Analysis of area under the ROC curve of energy detection," *IEEE Trans. on Wireless Commun.*, vol. 9, no. 3, pp. 1216-1225, March 2010.

[6]   H. Urkowitz, "Energy detection of Unknown deterministic Signals," *Proc. IEEE*, vol. 55, no. 4, pp. 523-531, April 1967.

[7]   F. Digham, M. Alouini, and M. Simon, "On the energy detection of unknown signals over fading channels," *IEEE Transaction on Communications*, vol. 55, no. 1, pp. 21-24, Jan 2007.

[8]   K. Hemachandra and N. Beaulieu, "Novel analysis for performance evaluation of energy detection of unknown deterministic signals using dual diversity," in *Proc. IEEE Veh. Technol. Conf.*, 2011, pp. 1-5.

[9]   S.P. Herath, N. Rajatheva, and C. Tellambura, "Unified Approach for Energy Detection of Unknown Deterministic Signal in Cognitive Radio Over Fading Channels," in *Proc. IEEE ICC Workshops*, 2009, pp. 1-5.

[10]   B. Modi, A. Annamalai, O. Olabiyi, and C. Palat, "Ergodic capacity analyses of cooperative amplify and forward relay networks over Rice and Nakagami fading channels," *International Journal of Wireless and Mobile Networks*, vol. 4, no. 1, pp. 97-116, Feb. 2012.

[11]   S. Alam, O. Odejide, O. Olabiyi, and A. Annamalai, "Further Results on Area under the ROC Curve of Energy Detectors over Generalized Fading Channels," in *Proc. 34th IEEE Sarnoff Symp.*, NJ, 2011.

[12]   K. Letaief and W. Zhang, "Cooperative communications for cognitive radio networks," *Proc. IEEE*, vol. 97, no. 5, pp. 878-893, May 2009.

[13]   F. Digham, M. Alouini, and M. Simon, "On the energy detection of unknown signals over fading channels," in *Proc. IEEE Int. Conf. Commun.*, vol. 5, 2003, pp. 3575-3579.

[14]   A. Liu, E. Schisterman, and C. Wu, "Nonparametric estimation and hypothesis testing on the

partial area under receiver operating," *Commun. Statistics - Theory and Methods*, vol. 34, no. 9, pp. 2077-2088, Oct. 2005.

[15]   A. Annamalai, C. Tellambura, and J. Matyjas, "A new twist on the generalized Marcum Q-function QM (a, b) with fractional-order M and its applications," in *Proc. IEEE Con. Commun. and Networking. Conf.*, 2009, pp. 1-5.

[16]   A. Annamalai and C Tellambura, "An MGF-derivative based unified analysis of incoherent diversity reception of M-ary orthogonal signals over fading channels,"," in *Proc. 54th IEEE Veh. Technol. Conf.*, 2001, pp. 2404 - 2408.

[17]   S.P. Herath, N. Rajatheva, and C. Tellambura, "On the energy detection of unknown deterministic signal over Nakagami channels with selection combining," in *IEEE CCECE '09*, 2009, pp. 745-749.

[18]   M. Simon and M. Alouini, *Digital Communication over Fading Channels*, 2nd ed. New York, USA: Wiley, 2005.

[19]   I. Gradshteyn and I. Ryzhik, *Table of integrals, series and products*. San Diego, CA, USA: Academic, 2007.

[20]   G. L. Stuber, *Principles of Mobile Communication*, 2nd ed. Norwell, MA: Kluwer Academic, 2001.

[21]   P. M. Shankar, "Error rates in generalized shadowed fading channels," *Wireless Personal Communications*, vol. 28, no. 4, pp. 233–238 , Feb. 2004.

[22]   P. Bithas, N. Sagias, P. Mathiopoulos, G. Karagiannidis, and A. Rontogiannis, "On the performance analysis of digital communications over generalized-K fading channels," *IEEE Commun. Lett.*, vol. 10, no. 5, pp. 353–355, May 2006.

[23]   S. P. Herath and N. Rajatheva, "Analysis of equal gain combining in energy detection for cognitive radio over Nakagami channels," in *IEEE GLOBECOM* , 2008, pp. 1-5.

[24]   A. Annamalai and C. Tellambura, "A simple exponential integral representation of the generalized Marcum Q-function QM (a, b) for real-order M with applications," in *Proc. 54th IEEE MILCOM*, 2008, pp. 1-7.

6

Optimizing Power and Buffer Congestion on Wireless Sensor Nodes using CAP (Coordinated Adaptive Power) Management Technique

Gauri Joshi and Prabhat Ranjan

Dhirubhai Ambani Institute of Information and Communication Technology
Gandhinagar, India
gauri_joshi@daiict.ac.in , prabhat_ranjan@daiict.ac.in

ABSTRACT

Limited hardware capabilities and very limited battery power supply are the two main constraints that arise because of small size and low cost of the wireless sensor nodes. Power optimization is highly desired at all the levels in order to have a long lived Wireless Sensor Network (WSN). Prolonging the life span of the network is the prime focus in highly energy constrained wireless sensor networks. Sufficient number of active nodes can only ensure proper coverage of the sensing field and connectivity of the network. If large number of wireless sensor nodes get their batteries depleted over a short time span then it is not possible to maintain the network. In order to have long lived network it is mandatory to have long lived sensor nodes and hence power optimization at node level becomes equally important as power optimization at network level. In this paper need for a dynamically adaptive sensor node is signified in order to optimize power at individual nodes along with the reduction in data loss due to buffer congestion.

We have analyzed a sensor node with fixed service rate (processing rate and transmission rate) and a sensor node with variable service rates for its power consumption and data loss in small sized buffers under varying traffic (workload) conditions. For variable processing rate Dynamic Voltage Frequency Scaling (DVFS) technique is considered and for variable transmission rate Dynamic Modulation Scaling (DMS) technique is considered. Comparing the results of a dynamically adaptive sensor node with that of a fixed service rate sensor node shows improvement in the lifetime of node as well as reduction in the data loss due to buffer congestion. Further we have tried to coordinate the service rates of computation unit and communication unit on a sensor node which give rise to Coordinated Adaptive Power (CAP) management. The main objective of CAP Management is to save the power during normal periods and reduce the data loss due to buffer congestion (overflow) during catastrophic periods. With CAP management we are trying to adaptively change the power consumption of sensor nodes. Power consumption of processing unit and communication unit are coordinated together and changed adaptively with respect to the workload. Coordination between processing and communication subunits results in better energy optimization as well as possible data loss before transmission because of limited buffer sizes can be avoided.

KEYWORDS

Power Optimization, Buffer Overflow, Wireless Sensor Nodes, Coordination between DVFS and DMS

1. UNDERSTANDING THE PROBLEM AND LITERATURE REVIEW

Micro sensor nodes are small in size and hence the size of battery supported on node is also small. This battery can supply a small amount of energy for the functioning of node and nodes are expected to work reliably over a longer period of time which may extend up to few years. Replacing the batteries is not feasible due to remote, random, inaccessible nodes and also due to large number of nodes in the network. Recharging or replacing the batteries on sensor nodes manually is not possible due to very large number of sensor nodes and also nodes may not be approachable physically in many applications. Limited battery energy constraint can be overcome if battery gets recharged using energy scavenging techniques. Electrical energy can be scavenged or harvested from environmental sources such as solar, wind or water flow. Continuous research [1-4] is going on to develop Energy harvesting WSN as it seems to overcome the problem of stringent power constraint. However for WSN applications the energy demands are large because of the wireless communications and availability of environmental power cannot be guaranteed as nodes energy harvesting opportunities may vary from place to place and time to time. Also the energy consumption of the nodes varies due to uneven distribution of workloads or network traffic. All these constraints emphasize that sensor node with energy scavenging ability also need power management.

Our focus is on the power optimization at sensor node level with ultimate aim to increase the lifetime of the Wireless Sensor Network. In a wireless sensor node maximum power is consumed for wireless communication and data processing also consumes moderate amount of power. We are trying to optimize the power consumption of processor and radio unit dynamically by adopting the optimized service rates (processing rate and transmission rate respectively) with respect to the instantaneous workload requirements. For a long-lived Wireless Sensor Network, low power hardware is the basic requirement [5] along with energy efficient protocols [6]. To increase the lifetime of wireless networks in which nodes are battery operated, various energy management techniques have been proposed in the literature. An overview of these techniques has been given in [7]. Mostly researchers have suggested reducing the wireless communication power [8, 9]. In [10-12]] shutting down processing unit has been suggested in order to optimize topology and communication range. Shutting down results in power saving. Information about network traffic and route is exploited in [13] to decide when to turn the node into the sleeping state from ON state. Though wireless communication is the major activity which consumes lot of power for transmission, reception as well as when it is idle (listening), power consumption of microcontroller or processor also have significant share in the energy budget of wireless sensor nodes. Emerging real time applications of WSN like body sensor networks need powerful microcontrollers which consume more power [14, 15]. In [16] energy consumption of the widely adopted Mica2 sensor node is studied very accurately and shown that the power consumption of processor ranges from 28 % to 86% of the total power consumed and roughly 50% on average. So it becomes equally important to optimize the power consumption of computing unit along with communication unit. Dynamic Frequency Scaling (DFS) and Dynamic Voltage Scaling (DVS) techniques are energy efficient when the processor is in idle state [17] to reduce power consumption. Topology control and power-aware routing protocols only reduce the transmission power of radio, and hence are not suitable for the applications with low workload or the radio platforms with high idle power consumption. Sleep scheduling protocols only reduce the idle power consumption and hence not effective when the network workload is high or the idle power consumption of radio is low. It clearly indicates the need for adaptive sensor node architecture which can handle both the situations of low traffic (normal period) as well as high traffic (catastrophic period) efficiently. That is we want no power wastage during normal periods and no data loss during catastrophic periods.

Figure 1 show the basic block schematic of a wireless sensor node.

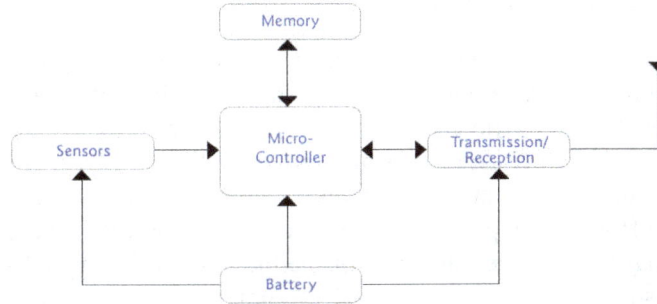

Figure 1. Basic architecture of a sensor node

Data arrives in the input buffer from two sources- data sensed by its own sensors and the data received from the neighboring nodes. The processor processes this data and the processed data comes in the output buffer. Transmitter transmits this data. Sensor nodes with fixed service rates are designed to handle moderate data arrival rates otherwise if designed to handle low data rates will result in data loss due to buffer overflow during catastrophic period or if designed to handle peak data rates then will remain idle over a longer period and power wastage will be more which reduces the lifetime of the sensor nodes.

Low duty cycling applied on low power hardware further increases the lifetime of the sensor nodes and long lived sensor nodes support to work the network over longer period. We have considered rotational sleep schedule (time triggered wake up) as it provides better network coverage and connectivity till sufficient number of nodes fails. As sensor networks are mainly deployed to sense some rare events, most of the times there is no much traffic in the network and not much work for the sensor nodes to carry. The problem arises when a node is turned ON but there is no much work to carry and hence remain idle for a longer duration. This idle state power consumption is the power wastage as power is consumed for doing nothing. More the idle period more is the more wastage. So it is important to control the power consumption during ON state by reducing the idle time periods. Here we have classified the time period over which sensor nodes are alive in the two categories- normal period and catastrophic period. Normal period is the time interval when event of interest has not occurred and everything is normal. It results in small data arrival rates in the input buffer of sensor nodes. Catastrophic period is the time duration when the event occurs. Lot of information is sensed by the nodes as well as lot of data received from the neighboring nodes results in the peak data arrival rate.

In order to reduce the idle power consumption during normal period if the service rates of the sensor nodes are kept smaller, then large amount of data arriving at the sensor node cannot be handled during the catastrophic period. Sensor nodes being very small sized devices have very small buffers to store the data and if not served with proper service rate then result in the buffer overflow and data gets lost. Hence wireless sensor nodes with longer life time but providing desired Quality of Service (QoS) are required.

Reducing the idle power consumption during the normal period and reducing the buffer overflow during the catastrophic period are equally important.

A sensor node capable of variable service rates can handle both the issues. Working with the smaller service rates consumes less power and reduction in idle power consumption increases the life time during normal period while offering higher service rates during the catastrophic period consumes more power but reduces the data loss because of buffer overflow.

We have modeled a sensor node with fixed service rates as well as a sensor node with variable service rates and compared their performances.

Data loss in a wireless communication system result by two ways. One possibility is that data is transmitted successfully and traveling on a communication channel data may be lost due to channel impairments, heavy traffic resulting in collisions, no connecting node etc. second possibility of data loss is before data transmission takes place. With a small sized hardware having constraints on its memory and buffer sizes, there is possibility of data loss due to buffer overflow. The tradeoff between packet loss due to buffer overflow and packet loss due to transmission errors has been studied in [18] for the increase in the overall system throughput. Data loss after transmission can be controlled to some extent by using proper modulation technique, channel encoding, error correcting codes etc. Data loss before transmission needs to take care by individual node. This kind of data loss is very prominent in the wireless communication networks where each node is highly hardware constrained as well as energy constrained. Wireless sensor network is a good example of this kind of network. Data loss in the node also results due to network congestion. Data is lost in the node itself before getting transmitted. Such loss occurs due to the buffer overflow at the output buffer (data loss due to buffer congestion). Data processed by processor comes in the output buffer and waits for getting transmitted. If the channel condition is poor then transmitter will not transmit the data in order to save the retransmissions. If receiver cannot receive the packet may be due to collision or something else then transmitter does not receive acknowledgement (ack) from the receiver and needs to keep the transmitted packet stored in the queue for retransmission. In such conditions probability of data loss due to output buffer congestion increases (either tail drop or head drop policy) as the output buffer size is very small in wireless sensor nodes. It not only leads to data loss but also results in CPU power wastage. It is wise to reduce the processing speed and save processors power if the transmission speed has reduced and packets in the output queue are waiting. In [14], effect of network congestion on buffer congestion has been analyzed and shown that by reducing the speed of microcontroller during congestion periods can save the power. We have discussed the effect of network congestion on buffer congestion resulting in data and power loss. Now consider the situation when there is no network congestion and channel condition is also good. For sensor network applications like monitoring applications most of the time there is no event occurring and there is very less traffic in the network. We call this time period as normal time period. When event occurs and get sensed and detected by the sensor nodes then the traffic in the network increases. We call this time period as catastrophic period. This increased traffic needs to be handled with proper processing and transmission speed otherwise data loss due to buffer overflow will occur. Information about the event of interest though sensed will not reach to the sink node. It may violet the purpose of deploying WSN. Figure 2 depicts the scope of Cap management technique for power and buffer overflow optimization at sensor node level.

2. POWER OPTIMIZATION AT SENSOR NODES

As discussed earlier data computation and wireless data communication are the main power consuming factors. By varying the computation (processing) and communication (transmission) rate w.r.t. the instantaneous amount of data to be processed or transmitted power can be optimized. Figure 2 elaborates the scope of coordinated power management technique.

2.1. Optimizing computing energy using DVFS

In most of the WSN applications sensor nodes have a time varying computational load, and hence peak system performance is not always required. DVFS exploits this fact by dynamically adapting the processor's supply voltage and operating frequency to satisfy the instantaneous processing requirement. The concept of dynamic voltage scaling is nicely elaborated by Amit Sinha, Anantha Chandrakasan et al [17, 20].

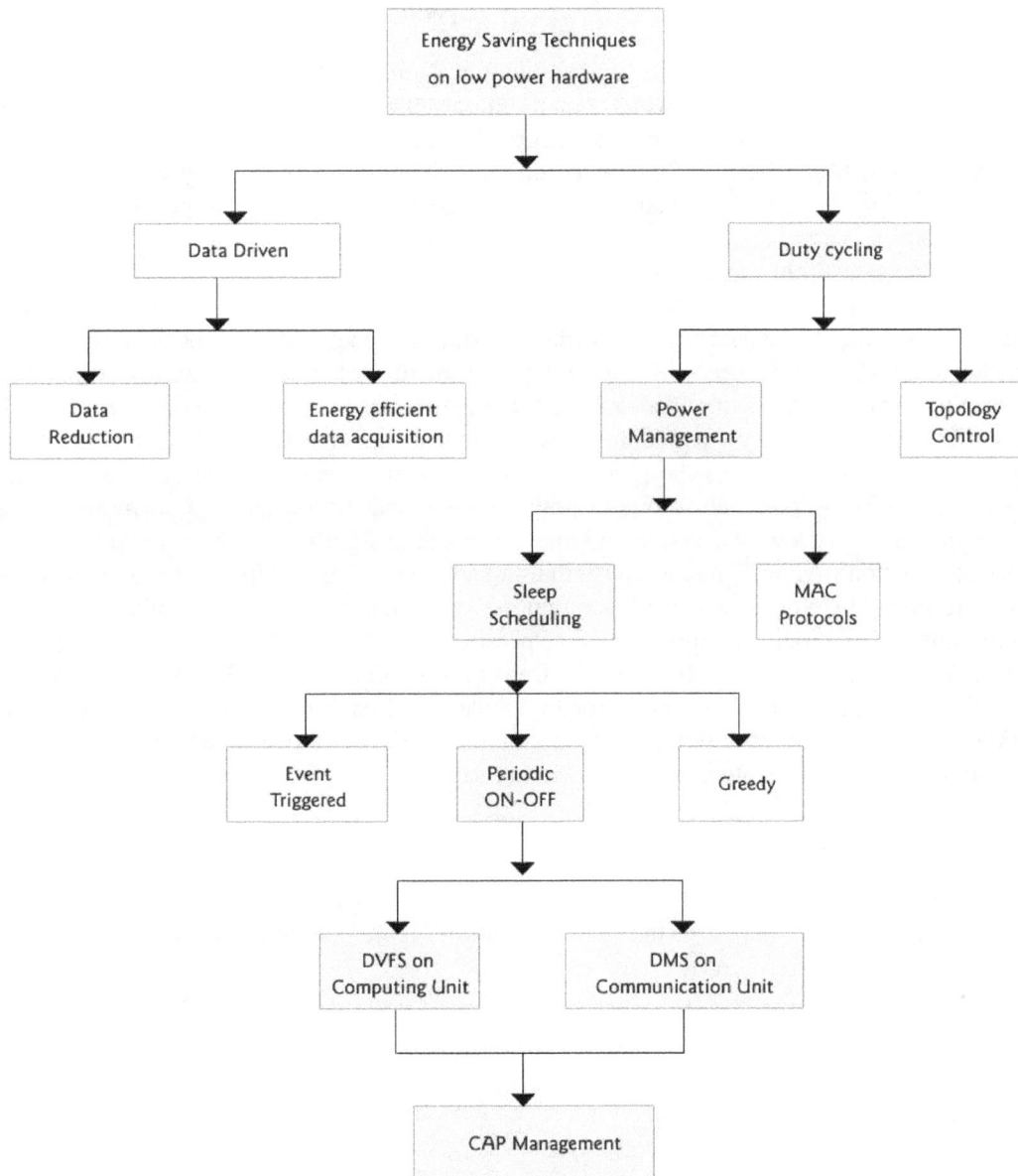

Figure 2. Scope of CAP Management on a Wireless Sensor Node

Here performance of processor is lowered against its energy efficiency. Performance is degraded in the sense that it takes more time for processing and introduces computational delay (latency), which is the cost paid to save computational energy. So this mean of computational energy saving can be adapted only within the latency constraint, which is not going to adversely affect the performance of the network. This latency constraint is different for different applications. Several modern processors such as Intel's Strong-Arm and Transmitta's Crusoe support scaling of voltage and frequency.

Reduction in the operational clock frequency results in linear energy savings and additional quadratic energy savings can be obtained if the power supply voltage is reduced to the minimum required for that particular frequency.

2.2. Optimizing communication energy using DMS

 M-ary modulation is the key to adaptive modulation. Number of bits per symbol (constellation size) can be changed adaptively which results in variable data rate but with constant symbol rate. In [21] it has been shown that MQAM modulation is efficient for short range communications. Modulation scaling is a technique to decrease the energy consumed during data transmission. Actual data transmission itself constitutes a major portion of the total energy consumption in wireless communication systems. Modulation scaling trades off energy consumption against transmission delay (latency).

Dynamic modulation scaling (DMS) concept is elaborated by Schurgers et. al [22]. It seems better to reduce the transmission time in order to reduce the energy consumption, so generally it is better to transmit as fast as possible and then turn to OFF state. Hence it is desirable to transmit multiple bits per symbol (M-ary modulation) in order to reduce on time of transmitter. But unfortunately for today's available transceivers start up time is much higher hundreds of microseconds) and it increases the power consumed by electronic hardware of the transmitter very aggressively as compared to output power transmitted. So switching transmitter ON and OFF frequently is not a wise decision and may not result in significant energy saving. In case of M-ary modulation ($M = 2^b$) as the constellation size b (number of bits per symbol) increases, power consumed by hardware as well as output power increases so for a particular transmission system value of b should be optimized for specific symbol rate. The energy consumption in data transmission is proportional to the transmission data rate [23, 24]. Increasing the constellation size b, energy consumed for transmission of each bit increases while associated transmission delay is decreased. Dynamic modulation scaling is useful to achieve multiple data rate and dynamic power scaling to provide energy savings.

3. SYSTEM MODEL

From the architecture of a sensor node it can be viewed as two systems connected in tandem (output of first system is input for the second system). Figure 3 shows the tandem queue model of a sensor node with fixed service rate.

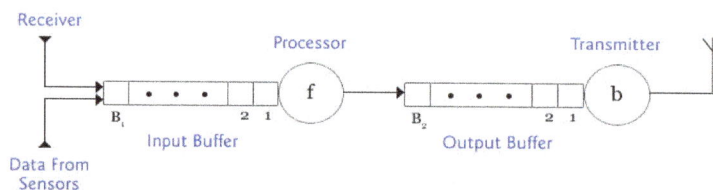

Figure 3. Tandem queue model of wireless sensor node (fixed service rate)

Let A_n = number of packets arrived during n^{th} slot

B_1 = maximum capacity of input buffer

B_2 = maximum capacity of output buffer

f = number of packets served by the processor in each time slot

b = number of packets transmitted by the transmitter in each time slot

M_n = input buffer occupancy at the start of nth slot

N_n = output buffer occupancy at the start of nth slot

We have considered late arrival system (LAS) where data packets are allowed to enter in the system just before the slot ends. These packets get service in the next time slot. Input buffer occupancy at the start of a time slot is dependent on buffer occupancy at the start of previous slot, number of packets served and number of packets arrived during previous slot.

$$M_n = \min \{\max \{(M_{n-1} - f), 0\} + A_{n-1}, B_1\}$$

Similarly output buffer occupancy can be written as

$$N_n = \min \{\max \{(N_{n-1} - b), 0\} + \min \{M_{n-1}, f\}, B_2\}$$

3.1. Fixed service rates (No DVFS, No DMS)

In most of the WSN applications during normal periods very less workload needs to be handled. When the number of packets in the buffer is less than the number of packets that can be served in one time slot of duration Δ, server remains idle for some period.

I_{1n} = Idle period of processor in n^{th} time slot

$$= \max \{(f - M_{n-1})/f, 0\} \Delta$$

Similarly,

I_{2n} = Idle period of transceiver in n^{th} time slot

$$= \max \{(b - N_{n-1})/b, 0\} \Delta$$

For a fixed service rate sensor node, M_n is a function of service rate arrival rate. As service rate is fixed M_n depends on arrival rate only. During normal period, arrival rate is very small which keeps the value of M_n also small and as a result processor remains idle over a longer period.

On the contrary during catastrophic conditions the arrival (number of packets arrived in one slot) increases but as the service rate is fixed and buffer size is small possibility of data loss due to buffer congestion (buffer overflow) occurs as per the head drop or tail drop scheme.

Input buffer overflow (OV_1) occurs when $M_n = B_1$ and output buffer overflow (OV_2) occurs when $N_n = B_2$.

$$OV_{1n} = max \{(M_{n-1} - f), 0\} + A_{n-1} - B_1$$

$$OV_{2n} = max \{(N_{n-1} - b), 0\} + min \{M_{n-1}, f\} - B_2$$

3.2. Only DVFS, No DMS (variable f and fixed b)

In this case input buffer status depends on arrival rate as well as service rate f. Arrival rate cannot be controlled but service rate can be used as a control knob to reduce idle period during normal times and buffer overflow during catastrophic period. In this case

$$M_n = min \{max \{(M_{n-1} - f_{n-1}), 0\} + A_{n-1}, B_1\}$$

This value of M_n will decide the value of service rate f_n. Smaller the value of M_n smaller value of f_n will be selected. Reduction in service rate will reduce the power consumption and will take more time to complete the service (DVFS). It helps to reduce the idle power wastage.

$I_{1n} =$ Idle period of processor in n^{th} time slot

$$= max \{(f_n - M_{n-1})/f_{max}, 0\} \Delta$$

By reducing the value of f_n idle time will be reduced and will also save the power.

Buffer overflow is given as

$$OV_{1n} = max \{(M_{n-1} - f_{n-1}), 0\} + A_{n-1} - B_1$$

During catastrophic condition as the arrival rate increases value of M_n will be more. Data loss due to input buffer overflow can be reduced by increasing the value of f_n. In order to make service rate buffer adaptive we need to scale f_n in terms of M_n.

$$f_n = (M_{n*} f_{max})/B_1$$

f_{max} is the maximum supported service rate.

But as the second server in the tandem queue (transmitter) works with fixed service rate, there is possibility of data loss during catastrophe and more power wastage during idle period. Output buffer occupancy is-

$$N_n = min \{max \{(N_{n-1} - b), 0\} + min \{M_{n-1}, f_n\}, B_2\}$$

Output buffer overflow can be written as

$$OV_{2n} = max \{(N_{n-1} - b), 0\} + min \{M_{n-1}, f_n\} - B_2$$

In this equation f_n is varying at the first server but there is no control knob at the second server to control the overflow. During catastrophe as A_n increases, fn at the first server will increase resulting in increased N_n. As b is constant and B_2 is fixed output buffer overflow increases it not only results in data loss but as the processed data gets lost, processing power used for that also goes waste. Similarly during normal conditions as A_n reduces, fn will be reduced. It will reduce the packet arrival rate in the output buffer but as second server works with fixed rate (which is high enough to handle worst case condition), it will remain idle over longer duration and more power will be wasted.

I_{n2} = Idle period of transceiver in n^{th} time slot

$$= \max \{(b - N_{n-1})/b, 0\} \Delta$$

N_{n-1} becomes smaller due to reduced f_n but b is constant and moderately high hence I_{2n} increases.

Implementation of only DVFS is not enough as it increases the processed data loss and processing power loss during catastrophe and more idle power wastage during normal period.

3.3. Only DMS, No DVFS (variable b and fixed f)

In this case service rate of processor- f is fixed and kept considerably high in order to handle sufficient number of packets during catastrophic conditions. Transmission rate b can be varied as per the number of packets in the output buffer. During normal conditions f will be much higher than the arrival rate A_n so the first server- processor remains idle over a longer duration. Idle power wastage is more likely in the first server. Number of packets entering in the output buffer is also small during normal conditions. Second server- transmitter will select lower service rate b for the transmission of packets. Lowering the transmission rate will reduce the power consumption (RF power with DMS). Similarly during catastrophic conditions A_n will increase and f is not sufficient high then data loss may occur at input buffer. At the output buffer data loss due to buffer congestion is reduced by increasing the transmission rate (more number of bits per symbol).

$$OV_{2n} = \max \{(N_{n-1} - b_n), 0\} + \min \{M_{n-1}, f\} - B_2$$

Here though b_n is changing, OV_{2n} gets limited by f which is fixed. So implementation of only DMS is also not enough.

3.4. Both DVFS and DMS integration (variable b and f)

From the discussions above it is highly desirable to have both f and b both changing w.r.t. to the number of packets in the buffer waiting for the service. Figure 4 shows the sensor node architecture with variable processing rate and variable transmission rate. Here a monitor checks the queue length and the probability of buffer overflow. Processing rate of the processor is varied as per the principle of dynamic voltage/frequency scaling (DVFS) and the data transmission rate is varied using dynamic modulation scaling (DMS).

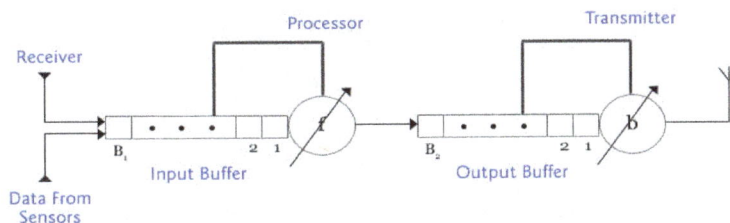

Figure 4. A sensor node with variable service rates

As seen earlier input and output buffer occupancies can be given as

$$M n = f \{A_n, f\} \qquad\qquad 0 \leq M n \leq B_1$$

$$N n = f \{f, b\} \qquad\qquad 0 \leq N n \leq B_2$$

for the stability of the system $f \geq A_n$, so that departure rate of first server is nothing but its arrival rate An. so we can approximate,

$$N n \sim f \{A_n, b\} \qquad\qquad 0 \leq N n \leq B_2$$

It shows that occupancy of input buffer as well as output buffer is a function of arrival rate A_n. Implementing DVFS (on processor) and DMS (on transmitter) together on a sensor node make the service rates f and b to change w.r.t. input and output buffer occupancy respectively will save the power during normal periods and will reduce buffer overflow data loss during catastrophic periods. Also as both the buffer occupancies are function of arrival rate A_n (directly proportional) there is no need to monitor input and output buffers separately. By monitoring input buffer only it is possible to select required f and b. Now we can say that f and b are changing in coordination. It results in coordinated adaptive power (CAP) management giving extended lifetime to the sensor nodes and indirectly contributing to the lifetime extension of WSN. It also ensures QoS by reducing the buffer congestion and data loss because of it.

4. NEED FOR CAP MANAGEMENT

In order to save the overall energy consumption of a sensor node we consider implementing DVFS and DMS on a sensor node. ON time period is divided in number of time slots of fixed duration Δ. At the start of every time slot status of the input buffer and out buffer are checked independently and accordingly voltage-frequency settings of the processor and modulation index of transmitter are set. Here as DVFS and DMS work independently on the same sensor node some of the problems that arise are:

- Two different power managers are required one for processing unit and other for communication unit.

- Input buffers and output buffer are separately monitored hence more interfacing with hardware is required.

- More software, more iterations and more energy required.

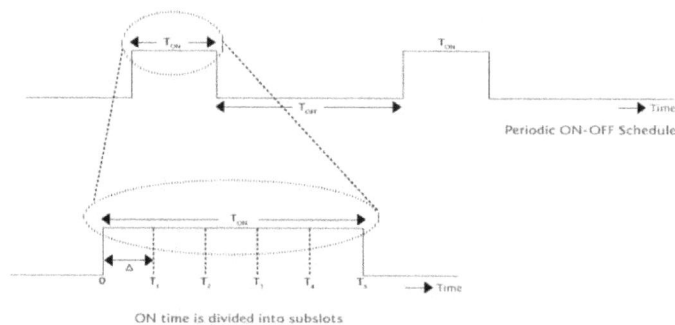

Figure 5. ON period divided in number of slots

- DVFS has predictive approach but DMS checks the actual status of the output buffer, so DMS monitor is not aware of what will be the status of output buffer in next slot.

- Most of the time sensor node simply has to forward the data, it does not consume any time for processing and suddenly the data in the output buffer increases.

As DMS modulator is not aware of this fact in advance so monitoring the output buffer, deciding and then adjusting the modulation level for transmission in the same time slot takes time. Meanwhile there is possibility of data loss due to limited buffer size.

To overcome above mentioned problems coordination between DVFS and DMS is required. If the operating state of communication unit is selected based on the operating state of the processor then both units will work together for power optimization as well as possibility of data loss before transmission gets removed.

4.1. Concept of CAP management

Idea of coordinated power management was floated by Vijay Raghunathan et al [24]. CAP management is a technique to coordinate active operating states of processor and transmitter in a particular time slot adaptively with workload. Dynamic Voltage Scaling (DVS) [25] and Dynamic Modulation Scaling (DMS) [26] techniques are integrated on the node and works in coordination. CAP management considers the following assumptions:

- Due to limited energy availability, a short haul multi hop communication is preferred.

- Other than sensing its environment each node acts as a router and simply forwards the received data to other nodes.

- Percentage of data to be forwarded is much greater than the percentage of data actually sensed.

- Predicted workload tracks the actual workload efficiently.

Using the fact that if workload monitor observes a heavy workload and selects higher supply voltage and clock frequency for processing predicted heavy load, then packet arrival rate in output buffer will be quite high and if packets are not transmitted quickly then some of the data packets may get lost even before transmission due to limited output buffer size. In this situation modulation scaler selects higher constellation size i.e. selects multiple bits per symbol and fast data transmission takes place. Transmission time Ton is reduced which results in to energy saving but electronic energy consumption for higher b increases which also depends on transmitter hardware and design. If DVFS is selecting smaller value of supply voltage and clock frequency then optimum constellation size b is selected, which results in to comparatively slow data transmission but reduces power consumed by hardware and also reduces output power transmitted. in wireless sensor networks for short haul communication generally 0dBm output power is considered which is much smaller than electronic power consumption. Hence it is not always worthy to transmit data with higher constellation size, but constellation size is optimized for minimum energy consumption and maximum latency constraints, rather better to scale constellation size with workload up to permissible limit.

In the block diagram shown in Figure 6, a common workload monitor and predictor controls voltage and frequency scaler for processor and also modulation (Mod) scalar for transmitter. As workload is predicted for next slot, where first data will be processed with certain processing rate and then will be made available in output buffer for transmission so a sufficient delay is introduced before giving control signal to modulator. Also constellation size used for modulation is required to be specified in the packet header for the purpose of demodulation at the receiver side.

Use of DVFS and DMS together has been explored in [27] and [28] for minimizing energy consumption. In [27], Kumar et al. addressed a resource allocation problem. In [28], genetic algorithm is used to solve the convex optimization problem of the energy management. In [29],[30] authors have combined DVFS and DMS techniques to maximize the battery energy levels of individual nodes at the same time meeting the end to end latency requirements.

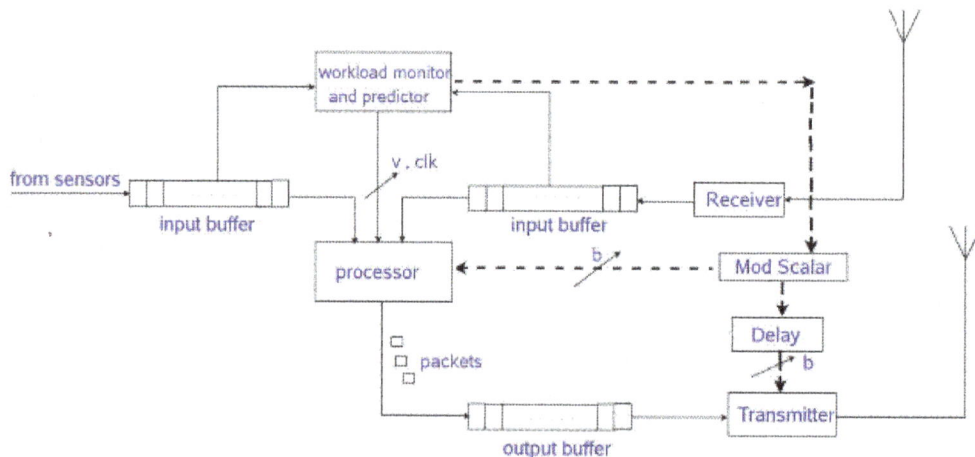

Figure 6. CAP Management Conceptual Block Diagram

5. SIMULATION RESULTS

We have assumed that the arrival of data packets follows Poisson distribution. During normal conditions the arrival rate A_n is assumed to be $\lambda 1$ while during catastrophe it is assumed to be $\lambda 2$. Both the system queues are of fixed lengths B_1 and B_2 respectively. When there is a sudden change in the surroundings the data flow increases to a comparatively higher rate and that is why there is a need for higher value of service rate during catastrophe period.

The node is designed in such a way that it analyses the overflow probability for both the servers after every 20 time units. During this period if the overflow of packets in any of the queues has reached above a threshold level then their respective service rates will increase, so that the higher traffic of data could be managed with less overflow.

As soon as the condition is back to normal the service rates will be changed back to the initial values. Using the lower values of service rates during normal conditions (because there is less data traffic in the system during normal conditions) we are trying to save the battery power since power consumption increases with the increase in service rate. And by increasing the value of service rates during catastrophe we are reducing the overflow of packets since during catastrophe data traffic in the system increases to a great extent.

Figure 7 and Figure 8 shows the graphs of power consumed, queue length and buffer overflow probability of a sensor node with fixed service rate and that of a sensor node with variable service rate. Figure 7 compares both types of sensor nodes under normal condition while Figure 8 shows comparison under catastrophic conditions.

In Table 1 all the performance parameters observed in both the models under normal as well as under catastrophe conditions are listed for the purpose of comparison. The first row depicts parameters obtained for the fixed service rate model. The rest of the rows depict parameters obtained for varying service rate model. Note that the values of varying service rate model show better results.

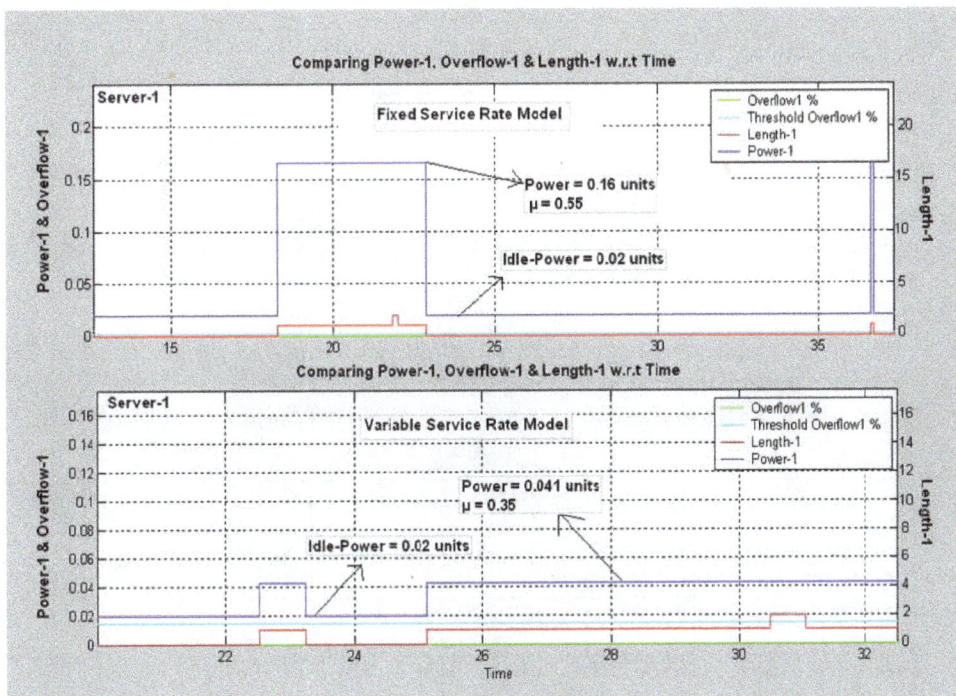

Figure 7. Performance graphs of processor with input buffer with fixed service rate and with variable service rates under normal condition

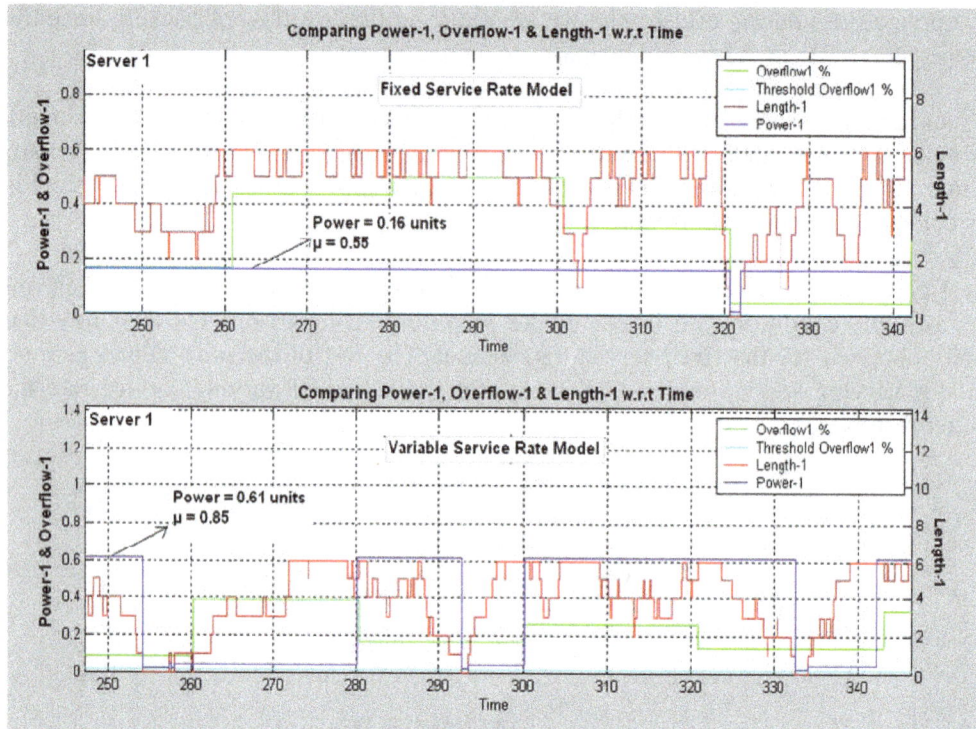

Figure 8. Performance graphs of processor with input buffer with fixed service rate and with variable service rates under catastrophe

Service Rate	f1,b1 (Normal)	f2,b2 (Catastrophe)	Ov1,Ov2 (Catastrophe)	Ov1,Ov2 (Normal)	P1,P2 (Catastrophe)	P1,P2 (Normal)	Idle time Probability	Lifetime
Fixed→	0.55,0.45	0.55,0.45	0.44 ,0.24	0.002,0.002	0.16,0.08	0.06,0.03	0.64,0.58	34620
	0.3,0.3	0.8,0.8	0.32,0.14	0.03,0.02	0.41,0.35	0.032,0.032	0.37,0.40	48260
	0.3,0.3	0.9,0.9	0.30,0.17	0.02,0.01	0.44,0.33	0.032,0.0029	0.36,0.40	44083
	0.3,0.3	1,0.8	0.24,0.13	0.02,0.02	0.61,0.33	0.033,0.027	0.36,0.40	39483
Variable	0.35,0.35	0.85,0.85	0.28,0.16	0.01,0.01	0.39,0.35	0.032,0.030	0.45,0.47	44044
	0.4,0.4	0.9,0.9	0.23,0.16	0.01,0.04	0.43,0.42	0.034,0.035	0.48,0.50	38504
	0.4,0.4	1,0.9	0.21,0.13	0.01,0.00	0.66,0.46	0.037,0.037	0.48,0.50	31401

Table 1: Performance parameters observed by simulation

f_1- f_2: service rates during normal (f_1) and catastrophic (f_2) period in server1; b_1- b_2: service rates during normal(b_1) and catastrophic(b_2) period in server2; Ov1-Ov2: Overflow probabilities in server1(Ov1) and server2(Ov2); P1-P2: Average power of server1(P1) and server2(P2); Idle time Probability: Probability that server1 and server2 are in idle state; Lifetime: Total lifetime of the node for the given energy.

6. CONCLUSION

A wireless sensor node with capability of adaptive service rates is more power optimized as compared with the sensor node with fixed service rate. Adaptive sensor nodes not only result in longer lifetime but also provide the better QoS by reducing the data loss due to the buffer overflow during the period of catastrophe. Longer lifetime is achieved by reducing the idle time periods and keeping sensor node busy with small service rates and consuming less power during normal period of operation. Service rate adaptive sensor nodes are actually power adaptive sensor nodes. These nodes also help to meet out the node-to-node delay constraints and reduce the number of time out dropped packets. Such long lived sensor nodes with better QoS performance will be helpful in making Wireless Sensor Networks more feasible.

In this paper we have considered only two ON states of a sensor node for the purpose of analysis. Similarly a sensor node with multiple number of states can be analyzed and will result in better performance as switching between two neighboring states will take less switching time and will consume less switching energy. Coordination between service rates of processor and transmitter reduces the need for checking two buffers independently. It gives a sensor node with coordinated DVFS and DMS techniques with better power optimization.

ACKNOWLEDGEMENTS

The authors would like to thank Prof. Jaideep Mulherkar, Mr. Sudhanshu Dwivedi and Mr. Anshul Goel for their interest and support in this work.

REFERENCES

[1]   Yu Gu, Ting Zhu, and Tian He. Esc: Energy synchronized communication in sustainable sensor networks, October, 2009.

[2]   Aman Kansal, Jason Hsu, Sadaf Zahedi, and Mani B. Srivastava. Power management in energy harvesting sensor networks. ACM Trans. Embed. Comput. Syst., 6, September 2007.

[3]   Shaobo Liu, Qinru Qiu, and Qing Wu. Energy aware dynamic voltage and frequency selection for real-time systems with energy harvesting. In Proceedings of the conference on Design, automation and test in Europe, DATE '08, pages 236–241, New York, NY, USA, 2008. ACM.

[4]   C. Moser, D. Brunelli, L. Thiele, and L. Benini. Real-time scheduling with regenerative energy. In In Proc. of the 18th Euromicro Conference on Real-Time Systems (ECRTS 06), pages 261–270. IEEE Computer Society Press, 2006.

[5]   Jason Hill, Mike Horton, Ralph Kling, and Lakshman Krishnamurthy. The platforms enabling wireless sensor networks. Commun. ACM, 47:41–46, June 2004.

[6]   Ian F. Akyildiz, Weiiian Su, Yogesh Sankarasubramaniam, and Erdal Cayirci. A survey on sensor networks. IEEE Communications Magazine, 40(8):102 – 114, August 2002. Survey.

[7]   Xiaodong Zhang Xiaoyan Cui and Yongkai Shang. Energy-saving strategies of wireless sensor networks. In Proceedings of the international Symposium on Microwave, Antenna, Propagation and EMC Technologies for Wireless Communications, 2007.

[8]   Ines Slama, Badii Jouaber, and Djamal Zeghlache. Optimal power management scheme for heterogeneous wireless sensor networks: Lifetime maximization under qos and energy constraints. In Proceedings of the Third International Conference on Networking and Services, pages 69–, Washington, DC, USA, 2007. IEEE Computer Society.

[9]   N. M. Moghadam A. S. Zahmati and B. Abolhassani. Epmplcs: An efficient power management protocol with limited cluster size for wireless sensor networks. In Proceedings of the 27th International Conference on Distributed Computing Systems Workshops, (ICDCSW'07), pages 69–72, 2007.

[10] Hung-Chin Jang and Hon-Chung Lee. Efficient energy management to prolong wireless sensor network lifetime. In Proceedings of ICI 2007. 3rd IEEE/IFIP Inter national Conference in Central Asia on Internet, 2007, pages 1 – 4, Sept. 2007.

[11] Y.-X. He C. Lin and N. Xiong. An energy-effcient dynamic power management in wireless sensor networks. In Proceedings of The Fifth International Symposium on Parallel and Distributed Computing, ISPDC'06, 2006.

[12] Xue Wang, Junjie Ma, and Sheng Wang. Collaborative deployment optimization and dynamic power management in wireless sensor networks. In Proceedings of the Fifth International Conference on Grid and Cooperative Computing, GCC '06, pages 121–128, Washington, DC, USA, 2006. IEEE Computer Society.

[13] D. Peng H. Wang, W. Wang and H. Sharif. Optimal power management scheme for hetero-geneous wireless sensor networks: Lifetime maximization under qos and energy constraints. In Proceedings of the International Conference on Communication Systems, ICCS, pages 1–5, 2006.

[14] Fabrizio Mulas, Andrea Acquaviva, Salvatore Carta, Gianni Fenu, Davide Quaglia and Franco Fummi. Network-adaptive management of computation energy in wireless sensor networks. In Proceedings of the 2010 ACM Symposium on Applied Computing, SAC '10, pages 756–763, New York, NY, USA, 2010. ACM.

[15] Mark Hempstead, Nikhil Tripathi, Patrick Mauro, Gu-Yeon Wei, and David Brooks. An ultra low power system architecture for sensor network applications. SIGARCH Comput. Archit. News, 33:208–219, May 2005.

[16] Victor Shnayder, Mark Hempstead, Bor-rong Chen, Geoff Werner Allen, and Matt Welsh. Simulating the power consumption of large-scale sensor network applications In Proceedings of the 2nd international conference on Embedded networked sensor systems, SenSys '04, pages 188–200, New York, NY, USA, 2004. ACM.

[17] A. Sinha and A. Chandrakasan. Dynamic power management in wireless sensor networks. Design & Test of Computers, IEEE, 18(2):62–74, August 2002.

[18] Anh Tuan Hoang and Mehul Motani. Cross-layer adaptive transmission with incomplete system state information. IEEE Transactions on Communications, 56(11):1961–1971, 2008.

[19] Fabrizio Mulas, Andrea Acquaviva, Salvatore Carta, Gianni Fenu, Davide Quaglia, and Franco Fummi. Network-adaptive management of computation energy in wireless sensor networks. In Proceedings of the 2010 ACM Symposium on Applied Computing, SAC '10, pages 756–763, New York, NY, USA, 2010. ACM.

[20] Rex Min, Travis Furrer, Anantha Chandrakasan, "Dynamic Voltage Scaling Techniques for Distributed Microsensor Networks," VLSI, IEEE Computer Society Workshop on, p. 43, IEEE Computer Society Annual Workshop on VLSI (WVLSI'00), 2000

[21] Zhang Jianhual Zhang Ping Shakya Mukesh, Muddassir Iqbal and Inam-Ur- Rehman. Comparative analysis of m-ary modulation techniques for wireless ad-hoc networks. In SAS, IEEE Sensors Applications Symposium,, February 2007.

[22] C. Schurgers, O. Aberthorne, and M. Srivastava, "Modulation scaling for energy aware communication systems," in *ISLPED '01: Proceedings of the 2001 international symposium on Low power electronics and design.* New York, NY, USA: ACM Press, 2001, pp. 96–99.

[23]   W. W. D. P. H. W. H. Sharif, "Study of an energy efficient multi rate scheme for wireless sensor network mac protocol," *Q2Winet06*, 2006.

[24]   V. Raghunathan, C. Schurgers, S. Park, and M. Srivastava, "Energy aware wireless microsensor networks," 2002

[25]   Hakan Aydin, Rami Melhem, Daniel Moss´,and Pedro Mej´a-Alvarez. Power aware scheduling for periodic real-time tasks. IEEE Trans. Comput., 53:584–600, May 2004.

[26]   Yang Yu, Bhaskar Krishnamachari, and Viktor K. Prasanna. Energy-latency trade- offs for data gathering in wireless sensor networks. In In IEEE Infocom, 2004.

[27]   G. Sudha Anil Kumar, Govindarasu Manimaran, and Zhengdao Wang. End-to- end energy management in networked real-time embedded systems. IEEE Trans. Parallel Distrib. Syst., 19(11):1498–1510, 2008.

[28]   Z. Fan C. Yeh and R.X. Gao. Energy-aware data acquisition in wireless sensor networks. In IEEE Instrumentation and Measurement Technology Conference, 2007.

[29]   Bo Zhang, Robert Simon, and Hakan Aydin. Energy management for time-critical energy harvesting wireless sensor networks. In SSS, pages 236–251, 2010.

[30]   B. Zhang, R. Simon, and H. Aydin. Joint Voltage and Modulation Scaling for Energy Harvesting Sensor Networks. *Proceedings of the First International Workshop on Energy Aware Design and Analysis of Cyber Physical Systems (WEA-CPS'10),* Stockholm, Sweden, April 2010.

7

Energy Efficient Scheduling Algorithm For S-MAC Protocol In Wireless Sensor Network

D Saha, M R Yousuf, and M A Matin

Department of Electrical Engineering and Computer Science, North South University, Dhaka, Bangladesh

m.a.matin@ieee.org

Abstract

In Sensor-MAC (S-MAC) protocol, a node located between two or more virtual clusters is called boarder node that adopts different listen and sleep schedules. These border nodes consume a large amount of energy as they switch to the listening mode often due to diversified scheduling which in turns decreases the lifetime of the wireless sensor network. This paper proposed a new unified scheduling method to solve the diversified scheduling problem of border nodes in S-MAC and evaluated the performance through simulation. It has been observed from the simulated results that the border nodes have consumed less power in case of large network as well as small networks.

Keywords

S-MAC protocol, Wireless Sensor Network (WSN), Uni-Scheduling packets

1. Introduction

Wireless sensor network (WSN) has become known as one of the potential applications of the emerging technology. It has been used for a wide range of purposes such as robotic exploration, environment monitoring, medical systems, etc. It has become a hot research issue which is regarded as one of the ten influencing technologies in the 21st century [1]. The WSN is made up of micro devices called Sensor Nodes (SN) which build the backbone of such networks and operate upon a predetermined set of instructions. As these sensor nodes are normally powered by batteries, they can only provide small and limited processing capabilities. The problem become very serious when these batteries are non-rechargeable in practical environment and a significant amount of power is spent for processing of required information. Thus a variant rate of depleted energy of the nodes can seriously hamper the network's efficiency and therefore its lifetime [2]. The major source of energy loss in WSN is *collision* or *corruption* during broadcasting of packets. The second major cause of energy loss is *idle listening*, that is, listening to an idle channel in order to receive any possible traffic. The next source of energy waste is *overhearing*, of a node to receive some packets that are destined for other nodes. The use of *control packet overhead* to setup data transmission is another source of energy loss. *Over emitting* is another source for energy waste which is caused by transmission of a message when the destination node is not ready.

Several type of protocols like TDMA and MAC [3] protocols are proposed to prolong the lifetime of WSN. In our paper, S-MAC (Sensor-MAC) protocol [4] is considered as it has shown better performance in minimizing power consumption. Yan-Xiao Li et al. [5] evaluated the performance of S-MAC protocol by improving latency and jitter. They have achieved their desired result by varying the duty cycle of the S-MAC. Though S-MAC protocol has shown better performance, it has some problems in initializing the listen and sleep schedules. To overcome this problem W. Lee et al. [6] designed an algorithm to minimize the energy consumption of border nodes by using unified schedule and modifying the synchronization packet. However, the nodes would incur delay and go through idle listening time if collision happened due to large packet size. E. M. Shakshaki et al. [7] presented a unified scheduling approach where the network got synchronized under a single schedule. Though the longevity of the WSN life has increased to some extent, the border nodes die out quickly as they stay in the listening state for longer period. T. S. Lee et al. [8] had shown the longevity of border nodes; but it is not suitable for large network size. Y. Yang et al. [9] performed scheduling from the sink node and therefore, in large networks the synchronization time will be longer and energy is wasted due to idle listening.

In this paper, an algorithm has been proposed to overcome the above limitations by using synchronization of sensor nodes under a unified schedule. This schedule will be broadcasted periodically from the border nodes. This means it broadcast constant synchronization packets to the whole sensor network, so that all the nodes can communicate with each other under a single schedule. In this way, the life time of all the nodes in the network will increase.

2. S-mac Protocol

S-MAC protocol is a medium access control protocol which is widely used in WSNs for energy conservation. It retains the flexibility of contention-based protocols similar to IEEE 802.11. There are three major energy consumptions in S-MAC that are identified. These are: a) *Collision* which results in energy waste due to retransmission of collided packets. b) *Overhearing* that occurs when a node listens to transmissions which are not intended for it. c) *Idle listening* which occurs when a node listens to receive any possible data that is not sent.

The communication between nodes takes place when S-MAC protocol exchange packets that starts with Carrier Sense (CS) to avoid collision. Then followed by Read to Send and Clear to Send (RTS/CTS) packets exchanged for unicast type data packets shown in the Fig-1. Upon successful transmission of these packets data communication takes place.

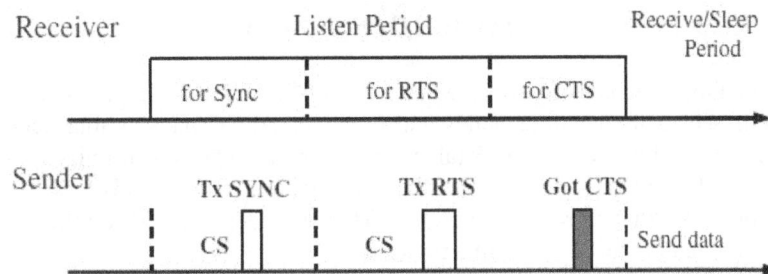

Fig.-1: S-MAC Messaging Mechanism

In [10], the nodes in WSN with S -MAC protocol keep on the listen state for 10 seconds in every 2 minutes. The node is not engaged in transmission or reception if S-MAC is in the sleep state or if its neighbors are involved in communication. The sleep state exits in S-MAC protocol to reduce collision and overhearing. The node wakes up at the end of its neighbor's transmission to relay the packets. This task is performed by overhearing neighbor's RTS and CTS exchanges and then the node goes to sleep and serve the purpose of reducing latency. This behavior of S-MAC protocol is called *adaptive listening* [11, 12] and the technique for optimizing power consumption is *message passing*.

Since all the immediate nodes have their own sleep schedules, periodic sleep may results in high latency. The latency caused by periodic sleep is called *sleep delay*. Schedules are periodically exchanged by broadcasting SYNC packets among neighboring nodes. The SYNC packet is very small and includes the address of the sender and the time of its next sleep. As soon as, a receiver gets the time from the SYNC packet it subtracts the packet transmission time and use the new value to adjust its timer. Due to the inconsistent of time cycle, different virtual clusters are formed. The communication between these virtual clusters takes place when a common node between them adopts both the schedules. In this way, the border nodes are listening for a longer period of time and die out quickly.

There are two main reasons of multiple scheduling in a single network. Firstly, when nodes establish their own schedules, some nodes are situated far away and cannot hear each other's schedules. Secondly, if two nodes broadcast their schedules at the same time, collisions may take place. In both the situations, the nodes must now choose their own schedules.

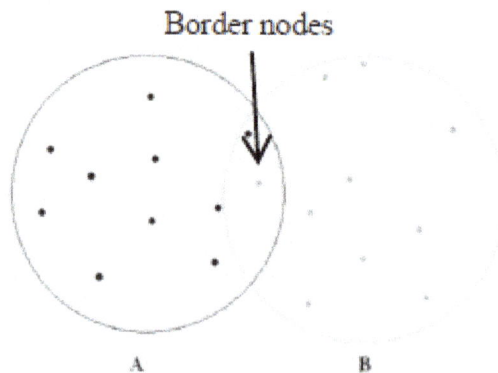

Fig.-2: Border node and Virtual Cluster

3. PROBLEMS WITH DIVERSIFIED SCHEDULING IN S-MAC PROTOCOL

3.1. DELAY TIME IN SYNCHRONIZATION OF SCHEDULE

S-MAC scheduling mechanism works when self-configuration is in set mode. In the listen period, a node senses its neighbor nodes and transmits SYNC packets that contain randomly generated schedule. Thus a long time is taken by each node to get synchronized. For instance, if 10 nodes are implemented in the network, they have to wait 100 seconds to setup the schedule and for 15 nodes the time rises to 150 seconds. Thus a longer time for stabilization takes place in proportion to the number of nodes in a network.

3.2. ISOLATED SCHEDULE CLUSTER

The virtual clusters are formed when some nodes have a common schedule. In case of individual cluster having one neighbor node or more is overlapped with another cluster of different period get to be synchronized, by receiving a SYNC packet. However in some cases the clusters can fail to acknowledge each other and no communication will take place between them, even though all the nodes are active. The S-MAC's solution to this problem is that all nodes work in the listen state for a given time.

In the actual implementation, all nodes sense SYNC packets and seek neighbor nodes while they are in the in the listen state for 10 seconds per 2 minutes. This phenomenon goes against the basic purpose of S-MAC, which states reduction of energy in the idle listen state. Thus a lot of energy is consumed while locating a hidden schedule cluster node.

3.3. DATA TRANSMISSION AFTER SYNCHRONIZATION OF BORDER NODE

The scheduling mechanism of S-MAC illustrates that each node has its own schedule generated randomly. It will then wait for a given time and if it fails to receive a SYNC packet, it will set its own schedule and broadcast its SYNC packet to the neighbor nodes. The neighbor nodes will receive the SYNC packet and use that schedule to get synchronized, but the whole network is not unified under same schedule. Therefore an independent schedule cluster having an independent schedule gets to be made and a node between heterogeneous schedules gets to receive SYNC packets which are different from one another and work as a border node [10].

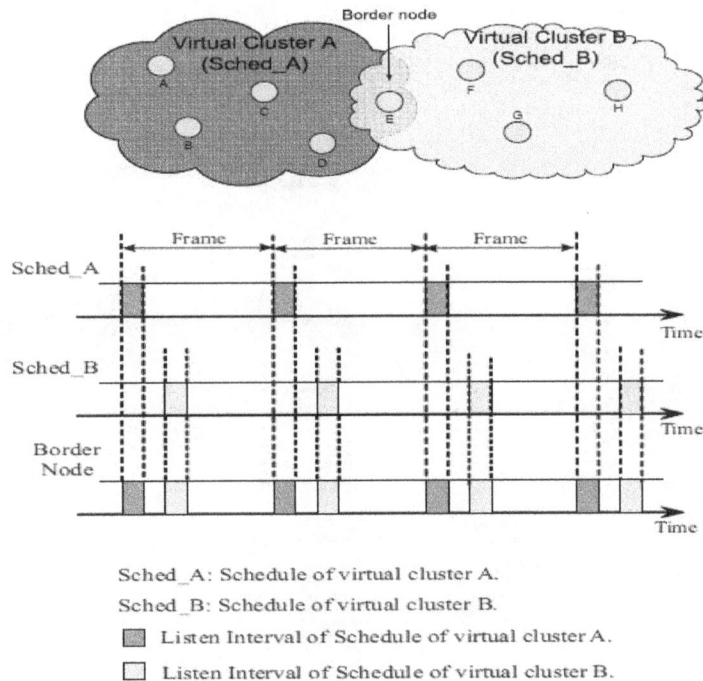

Fig.-3: Border Node adopting and handling both the schedules

As shown in Fig. 3, the border node adopts and handles both the schedules and creates a link between the virtual clusters. Therefore, it has to be in the listen state twice and the power consumption will be twice of a general node. If the border node dies out quickly, the clusters will not be able to interact with each other and no data transmission will take place between them. Thus the power consumption of the border node will increase in proportion to the number of different schedules adopted by it.

In the existing S- MAC code, this problem is somewhat minimized as the border node will only adopt one schedule depending upon the SYNC packet received first. However, this does not change the longevity of the border node as it stays in the listen state for a longer time.

3.4. ERROR DEVELOP DURING SYNCHRONIZATION

When the neighbor nodes synchronize their sleep schedule, the clock drift between each node can cause synchronization errors. This error takes place when SYNC packets are broadcasted periodically. At present, two techniques are in the S-MAC protocol code to minimize this error. First, all exchanged timestamps are relative rather than absolute. Second, the listen period is significantly longer than clock drift rates. The clock drift between two nodes does not exceed 0.2ms per second [10]. If we consider the listen time to be 0.5 second, as about 2500 second pass for a clock drift to take place. The listen time will increase the clock drift and consequently, increase the synchronization error. Thus, a time error resulting from a long clock drift is a critical factor in the actual implementation.

4. ALGORITHM DEVELOPMENT

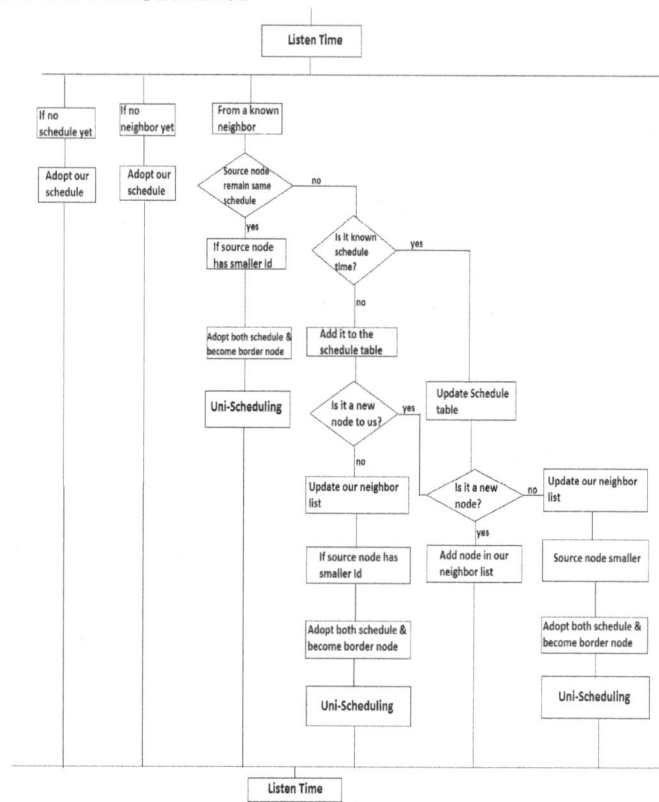

Fig.-4: Implementation of border node in the proposed algorithm

Our main focus will be on implementing a scheduling scheme so that all the nodes can communicate with each other at the same time. In order to synchronize the whole WSN quickly, our proposed algorithm is performed from the border nodes and more data packets can be transferred between the different virtual clusters

Firstly, if a node has its own schedule and receives a SYNC packet from a node with larger ID, it will not adopt the latter schedule. This is because, nodes with larger ID are newly joined in the network and will stay in the network for a shorter time. Secondly, the Neighbor list Table is updated after updating the Schedule Table in order to know the entire active neighbor surrounding a node. Thirdly, the broadcasting of *Uni-Scheduling packets* will be performed after the construction of border nodes. The construction of a border node will take place during the listening period and will perform the following tasks as shown in Fig.4. Fourthly, the whole synchronization takes place during the listening period.

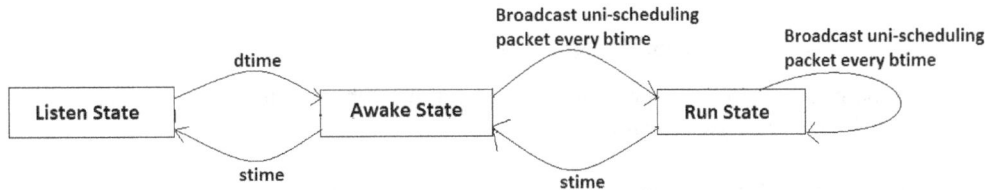

Fig.-5: State transition for Proposed Uni-Scheduling

After the border nodes are constructed, the next task is to set up a common schedule across the network so that all the nodes can be unified under a single schedule instead of heterogeneous schedules. The proposed unified scheduling algorithm is shown in Fig.5. Before going to analysis the proposed algorithm, several assumptions are taken which are as follows:

i.) All the nodes should be fixed without any mobility.
ii.) All the nodes have the same listen and sleep period throughout the simulation.
iii.) The adaptive listening is not applied for this simulation.
iv.) All the nodes are synchronized according to the schedule in the Uni-Scheduling packets.
v.) The *Uni-Scheduling packets* are broadcasted periodically from the border nodes.
vi.) If a node is running under uni-schedule and if it still receives any more *Uni-Scheduling packets*, it will discard all the latter packets.

The sensor nodes deal with *Uni-Scheduling packets* in the proposed algorithm and the transmission of these packets to all the nodes are done by using three types of timers. All the nodes go through three transition state before they are synchronized. The detail explanation of the different transition state and timers are given below

- The *LISTEN STATE* is the initial node state in which the node works in the listen state. The state is periodically changed to *AWAKE STATE* according to the *D-time*.
- In the AWAKE *STATE*, the border nodes adopt the uni-schedule as their primary schedule in the Schedule Table. Then *Uni-Scheduling packets* are broadcasted periodically to all the nodes in the network after every *B-time*. No data packets transmission is performed until the nodes receive *Uni-Scheduling packets*.
- After receiving *Uni-Scheduling packets* the nodes state change to the *RUN STATE*. In the *RUN STATE* the nodes will sense if their neighbors are synchronized under uni-schedule. After that data transmission will take place.
- The *D-timer* is used to change the transition state of the node periodically from the *LISTEN STATE* to the *AWAKE STATE*.
- The *B-timer* is used so that *Uni-Scheduling packets* are broadcasted periodically from the border nodes to all the other nodes in different virtual clusters.
- The *P-timer* is used to set a time of which the nodes waits to receive the subsequent *Uni-Scheduling packets* and if a node failed to receive any *Uni-Scheduling packets*, the node will move back to the previous state after the timer expire.

The *Uni-scheduling packet* as shown in Fig.6 contains a scheduling time and this packet is broadcasted from the border nodes to the whole sensor network. According to this signal, the nodes reset their timers and set the uni-schedule as their primary schedule in their Schedule Table. After that the nodes will be synchronized and data transmission begins.

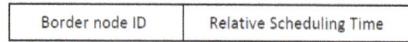

Border node ID	Relative Scheduling Time

Fig.-6: The structure of Uni-Scheduling Packet

When all the nodes follow this proposed uni-scheduling scheme, the whole network gets to be synchronized under a single schedule.

5. ADVANTAGE OF THE PROPOSED ALGORITHM

Using this proposed Uni-Scheduling algorithm, the nodes do not have to take longer period of time for synchronization. This is because all the nodes will run under a single schedule and will transmit data at the same time. Thus the delay time eliminated as stabilization of schedule has been solved through the synchronization of the network. There is no generation of isolated schedule cluster due to Uni-Scheduling process. Even if, a new node wants to add to the network, it can receive the *Uni-Scheduling packets* in the *AWAKE STATE* and quickly change to the *RUN STATE* and the network can be expanded easily without any problem. Moreover, all the nodes in the network are unified under the same schedule and the energy consumption in the idle state is reduced.

As soon as, the nodes get synchronized after receiving the *Uni-Scheduling packets,* the data can be transmitted and received continuously. Moreover, the conventional border node concept does not exist no more and works similar to other nodes. In this way the energy consumption of the sensor nodes are almost similar.

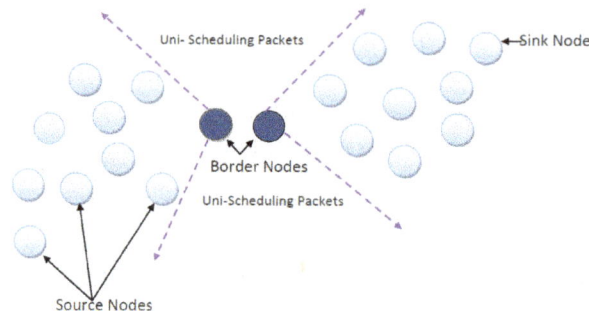

Fig.-7: Broadcasting *Uni-Scheduling Packets* to the network

When all the nodes get synchronized under a single schedule after receiving the *Uni-Scheduling packets,* the clock drift between two nodes are minimized. This clock drift error is reduced because all the nodes will perform their transmission at the same time. The synchronization error in the WSN is also reduced and this effect is clearly visible when the network size increases.

6. NETWORK SETUP

Table-1: PARAMETERS FOR SIMULATION

Parameter	Values
RTS/CTS/ACK Size	10 Bytes
Sync Packet Size	9 Bytes
Uni-Scheduling Packet Size	9 Bytes
Simulation area	1000m x 1000m
Routing Protocol	DSR
Listen Time	0.5 sec
D-Time, B-Time & P-Time	5 sec, 1sec & 200 sec
Simulation Time	1000 sec
Duty cycle	10%
Initial Energy	100 J
Tx and Rx Power	10 and 20

The proposed scheme is simulated in the Network Simulator-2 (ns-2) [13] under Linux environment. The platform for this simulation has provided us the perfect way to visualize the sensor nodes. All the necessary parameters for the wireless network and the energy model are included in the Tcl script. The TCP sink Agent is attached to the biggest node and TCP Agent is attached to all the remaining nodes in the network. The traffic application for this simulation is CBR traffic. Four kinds of topologies are considered for testing purpose and these are shown in Fig.8.

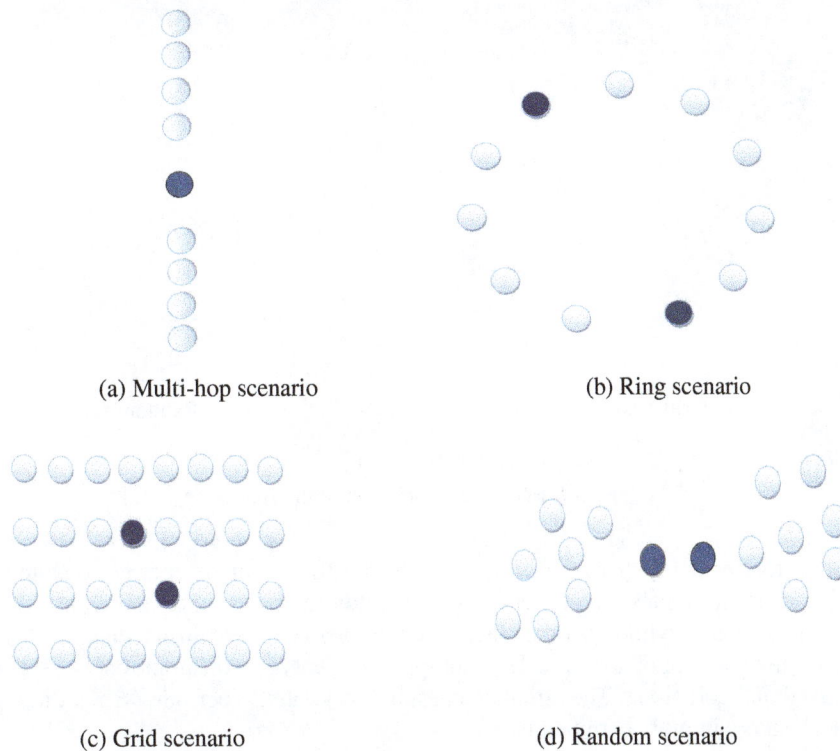

(a) Multi-hop scenario

(b) Ring scenario

(c) Grid scenario

(d) Random scenario

Fig.-8: Network Topologies

7. SIMULATED RESULTS AND ANALYSIS

In this section, the simulation results for different topologies are presented. The topologies that are used consist of multi-hop network scenario, ring network scenario, grid network scenario and random network scenario. The number of nodes used for first two network scenarios are 10 nodes and the last two scenarios consist of 40 nodes.

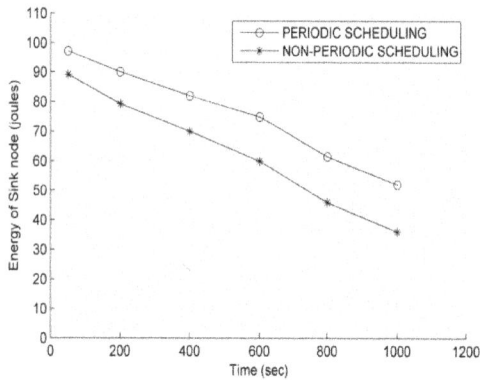

(a)   Multi-hop scenario

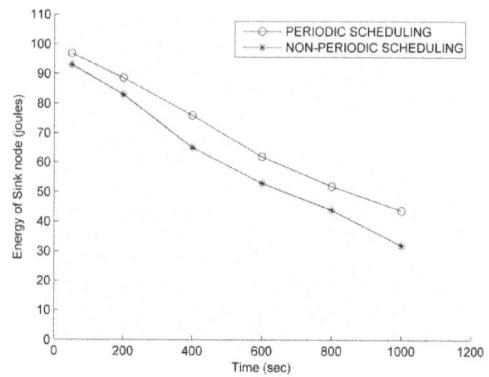

(b) Ring scenario

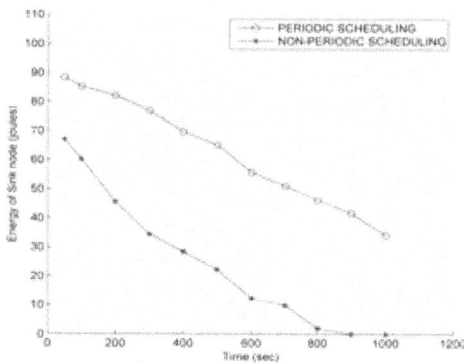

(c) Grid scenario

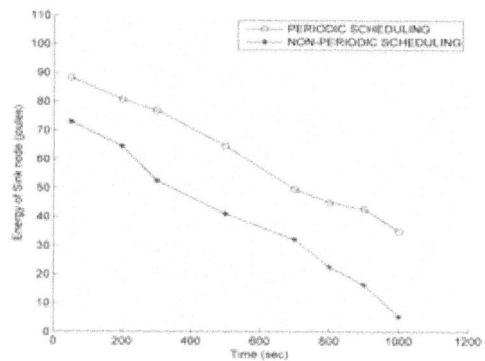

(d) Random scenario

Fig.-9: Energy consumption of the sink node

The above figures 9 (a), 9 (b), 9(c) and 9(d) show the result of energy consumption of the networks when the nodes adopt sleep/listen schedule and the other algorithms do not adopt sleep/listen schedule, while all other factors of the network remain unchanged. It is observed that if the network topologies are less complicated, energy consumption is less due to less overhearing and collisions. The simulated results prove that   periodic scheduling saves large amount of energy in more complex networks.

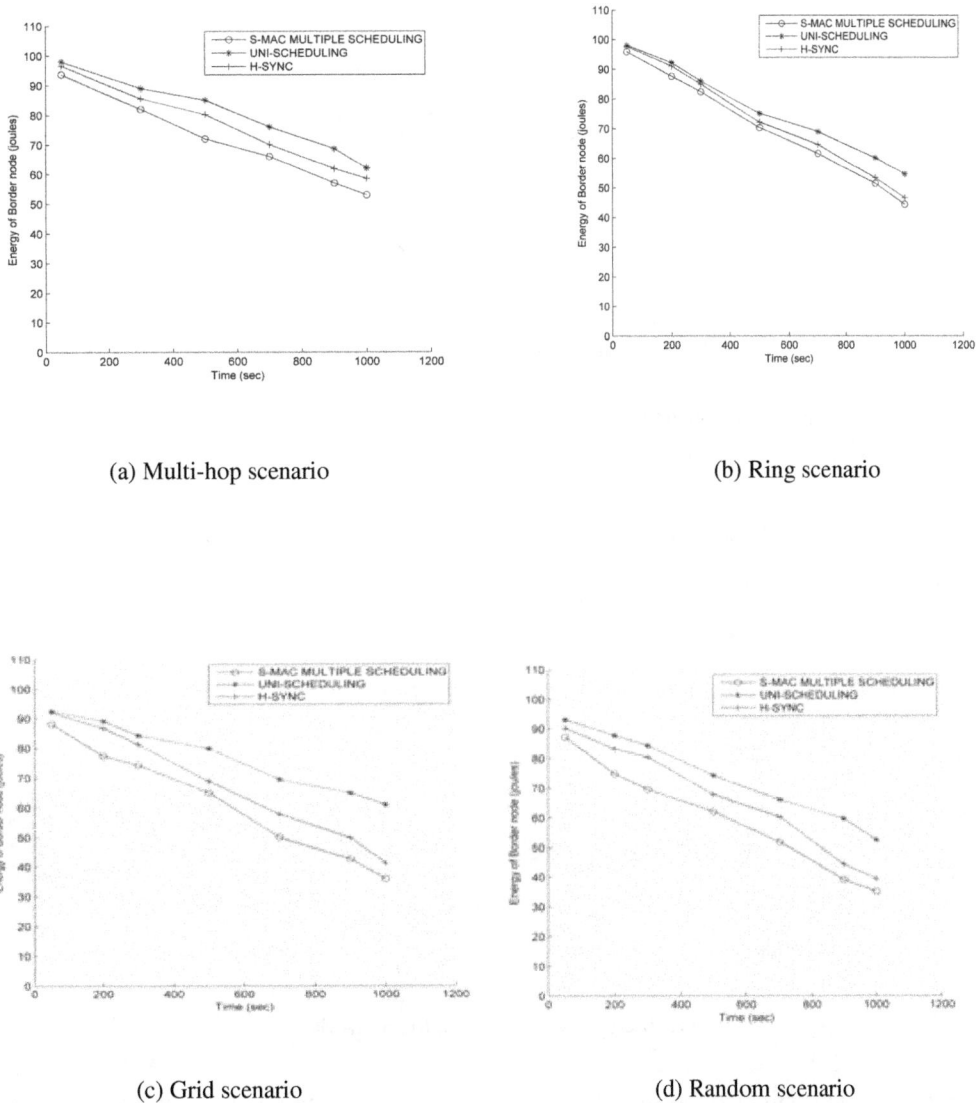

(a) Multi-hop scenario

(b) Ring scenario

(c) Grid scenario

(d) Random scenario

Fig.-10: Energy consumption of the border nodes

Fig.10 shows the energy consumption of the border nodes by applying different scheduling methods. In this experiment we have verified that our proposed Uni-Scheduling scheme performs more efficiently when compared with one of the unified scheduling methods called H-SYNC mention in paper [8] and S-MAC multiple scheduling. In smaller network topologies like the ring network, the energy consumption of all the scheduling methods is almost similar; no major changes can be distinguished from smaller network topologies. A significant result can be obtained from complicated networks, where the border nodes tend to lose more energy when they adopt H-SYNC and S-MAC scheduling. In the case of random scenario the difference in energy consumption between H-SYNC and S-MAC is 14.82% and between Uni-Scheduling and S-MAC is 69.67%. Thus we see that at the end of simulation Uni-scheduling saves 4.70 times more energy when compared with other scheduling methods. Therefore the life span of the border nodes increases and more data can be transferred between different virtual clusters.

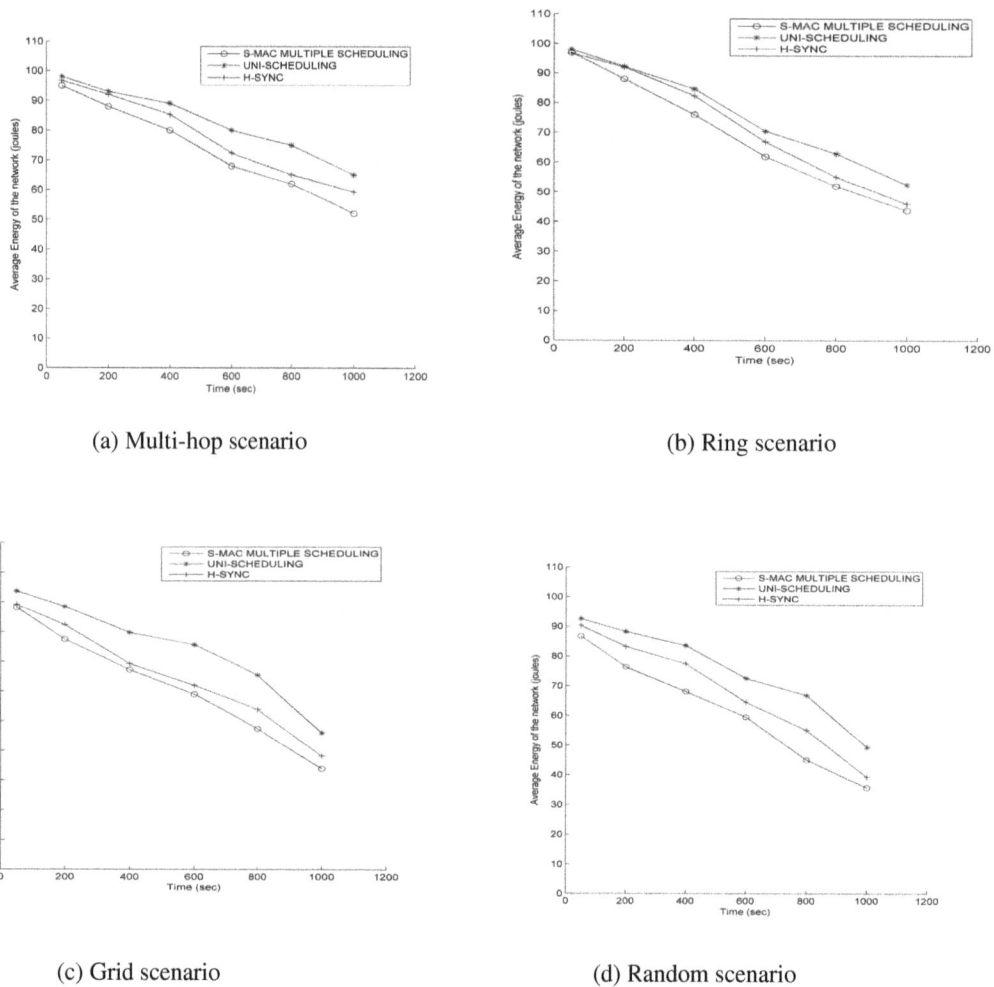

(a) Multi-hop scenario

(b) Ring scenario

(c) Grid scenario

(d) Random scenario

Fig.-11: Energy consumption of the network

From the previous simulated results shown in Fig.10, it has been noticed that the longevity of the border nodes for Uni-Scheduling are much longer than other scheduling methods. The graphs clearly show that as the network scenario changes, the energy consumption of the nodes increase significantly. In the case of grid and random scenarios, the energy consumption between Uni-Scheduling and H-SYNC is 11.91% and 18.73% respectively. Thus the Uni-Scheduling method is more appropriate than any other unified scheduling method. This also means that Uni-Scheduling is more appropriate for a complex WSNs.

8. CONCLUSION

The proposed scheme has unified the scheduling of S-MAC protocol and overcame the problems of diversified scheduling in wireless sensor networks. The border nodes in between virtual clusters broadcast Uni-Scheduling packets which synchronize the network under a single schedule. The simulated results showed that our proposed unified scheduling scheme performed much better than other scheduling methods. Thus, the lifetime of the network increases in compare to the existing S-MAC protocol.

REFERENCES

1. S. Wen-Miao, L.Yan-ming and Z. Shu-e, "Research on SMAC protocol for WSN" in Proc. IEEE 4th International conference on Wireless Communication, Networking and Mobile Computing WiCOM '08, Oct 2008, pp.1-4.

2. B Paul and M. A. Matin, "Optimal geometrical sink location estimation for two-tiered wireless sensor networks" IET Wirel. Sens. Syst., 2011, vol.1, no. 2,pp.74-84

3. Wireless LAN Medium Access Control (MAC) and Physical Layer (PHY) Specification, IEEE Std 802.11-1999 edition.

4. Wei Ye, John Heidemann and Deborah Estrin, "An Energy-Efficient MAC Protocol for Wireless Sensor Networks", in Proc. IEEE INFOCOM, New York, pp. 1567-1576, June 2002.

5. Yan-xiao Li, Hao-Shan Shi, and Shui-Ping Zhang, "An Energy-Efficient MAC protocol for Wireless Sensor Network", 3rd International Conference on Advanced Computer Theory and Engineering(ICACTE), Chengdu, Vol. 6, No. 4, pp. 619-623 Aug. 2010.

6. Woonsik Lee, Minh Viet Nguyen, Arabinda Verma, Student Member, IEEE, and Hwang Soo Lee, Member, IEEE, "Schedule Unifying Algorithm Extending Network Lifetime in S-MAC Based Wireless Sensor Networks", IEEE Transactions on Wireless Communication, Vol. 8, No. 9, pp. 4375-4379, Sept. 2009.

7. Elhadi M. Shakshuki, Tarek R. Sheltami, Haroon Malik, Chuoxian Yang, "Investigation and Implementation of Border Nodes in S-MAC", International Conference on Network-Based Information Systems, Indianapolis, Indiana, pp 350-357, Aug. 2009.

8. Tae-Seok Lee, Yuan Yang, Ki-Jeong Shin and Myong-Soon Park, "An Energy-Efficient Uni-Scheduling based on S-MAC in Wireless Sensor Network", pp 293-304, 2005.

9. Yuan Yang, Fu Zhen, Teo-Seok Lee and Myong-Soon Park, "TASL: A Traffic-Adapted Sleep/Listening MAC Protocol for Wireless Sensor Network", International Journal of Information Processing Systems, Vol.2, No.1, pp. 39-43, March 2006.

10. Wei Ye, John Heidemann and Deborah Estrin, "Medium Access Control With Coordinated Adaptive Sleeping for Wireless Sensor Networks", IEEE/ACM Transactions on Networking, VOL. 12, No. 3, pp. 493-506, June 2004.

11. Baoshan Zhang, Xiangdong Wang, Shujiang Li, Leishu Dong, "An Adaptive energy-efficient Medium Access Control Protocol For Wireless Sensor Networks", IEEE 2009 Fifth International Conference on Mobile Ad-hoc and Sensor Networks, Fujian, Wu Yi Mountain, China, Vol. 6, No. 9, pp. 350-357, Dec. 2009.

12. Behnam Dezfouli, Marjan Radi and Shukor Abd Razak, "A Cross-Layer Approach For Minimizing Interference And Latency Of Medium Access In Wireless Sensor Networks", International Journal of Computer Networks & Communication, Vol.2, No.4, pp. 126-142, July 2010

13. http://www.isi.edu/nsnam/ns/

8

VERIFICATION OF ENERGY EFFICIENT OPTIMIZED LINK STATE ROUTING PROTOCOL USING PETRI NET

Radhika D. Joshi [1] and Priti P.Rege [2]

[1]Department of Electronics and Telecommunication Engineering, College of Engineering, Pune, India
`rdj.extc@coep.ac.in`
[2]Department of Electronics and Telecommunication Engineering, College of Engineering, Pune, India
`ppr.extc@coep.ac.in`

ABSTRACT

Self-organizing ad hoc networks need development of efficient routing protocols in terms of reliable routing and energy conservation. For dense ad hoc networks, Optimized Link State Routing (OLSR) protocol is suitable due to its multipoint relaying (MPR) feature. We have tried to make the OLSR energy efficient by making effective neighbor selection based on residual battery energy of a node and traffic conditions that influence the drain rate of the node in the network. We have considered the multipath and source routing concept for route selection and a route recovery technique to tackle mobility issue efficiently. Modifications make the protocol energy efficient and at the same time achieve balancing of network load. Simulation results of OLSR and the modified protocol, show improvement in 'Number of nodes alive' against variation in pause time, speed, node density and simulation time. The work describes modified protocol verification using Petri net for future instances.

KEYWORDS

MANET, OLSR protocol, multipath routing, Petri Net

1. INTRODUCTION

Wireless ad hoc networks are autonomous, self-configuring and adaptive. Thus, such networks are excellent candidates for military tactical networks, where their ability to be operational rapidly and without any centralized entity is essential [1]. The development of multimedia service brings new opportunities and challenges to the wireless network technologies.

In an ad hoc network, each node creates a network link in a self-organizing manner, forwarding data packets for other nodes in the network [2]. Mobile ad hoc networks (MANETs) are instantly deployable without any wired base station or fixed infrastructure [3]. Due to these features, MANETs suffers from limitations like lower capacity, limited security, higher loss rates, more delays and jitter as compared to fixed networks. A critical issue for MANETs is that the activity of node is energy-constrained [4]. In MANET, operations of nodes rely on batteries or other exhaustible power supplies for their energy. Hence depletion of batteries will have greater effect on overall network. As a consequence, energy saving is an important system design criterion. Furthermore, nodes have to be power-aware: the set of functions offered by a node depends on its available power (CPU, memory, etc.) [5]. Significant energy savings can be obtained at the routing level by designing minimum energy routing protocols that take into consideration the energy costs of a route when choosing the appropriate route. The Multipath routing protocols consist of finding multiple routes between a source node and a destination node. These multiple paths can be used to compensate for the dynamic and unpredictable nature of ad hoc networks [6].

Though the major motivation of studying ad hoc networks comes from military usage, they will also be useful in several forms of tactical communication such as disaster recoveries, explorations, emergency services, law enforcements, educational applications, entertainment, location aware services and in various forms of home and personal area networks, as well as sensor networks [7, 8].

We present remainder of the paper as follows. In Section 2 we discuss issues of routing protocols and energy efficient routing protocol techniques in MANET. Original OLSR protocol features that are proposed to be modified are also discussed. Section 3 describes our scheme for making OLSR energy efficient. To modify existing protocol, various changes that are incorporated are illustrated in sub sections of Section 3. Section 4 includes simulation environment scenario used in NS-2 simulator. Section 5 shows performance comparison of OLSR and modified protocol, based on simulation results. The modified protocol verification using Petri net is illustrated in Section 6. Section 7 concludes the work focusing on, improvement in network behavior by using modified protocol, and verification of modified protocol using Petri net.

2. LITERATURE SURVEY

Routing protocols have to suggest best possible path from source to destination for efficient data transfer. For any application, the mobility of nodes as well as limited battery resources must be considered as design issues for expecting best performance from the network under consideration. It is very difficult to have correct data delivery under mobility conditions and to save the node power at the same time. Routing protocols are classified as Proactive, Reactive and Hybrid based on the method of maintaining route information in the protocol. In proactive protocols all routes are maintained regardless of the state of use [9].

Various techniques for making routing protocol energy efficient are considered. Saoucene Mahfoudh et al. [10] have distinguished three families of energy efficient routing protocols. Few proposals especially focused on the design of routing protocols providing efficient power utilization are dealt in depth by C.K.Toh [11]. The techniques are, Minimum Total Transmission Power Routing (MTPR), Minimum Battery Cost Routing (MBCR), Min-Max Battery Cost Routing (MMBCR), and Conditional Max- Min Battery Capacity Routing (CMMBCR). In addition to above techniques, minimum drain rate mechanism also needs to be considered for power saving. The drain rate is the rate at which energy gets dissipated at a given node. Each node monitors its energy consumption and maintains its battery power drain rate value during the given past interval.

Multipath routing offers several benefits like load balancing, fault-tolerance, higher aggregate bandwidth, lower end to end delay, reduced bottlenecks and security [12]. Multipath routing protocols have the advantage of sharing load of any flow on several paths, leading to a lesser consumption on the nodes of the selected paths. Multi-path routing techniques are proposed to use the minimized energy consumed per bit in discovering the least-required energy routing paths while reducing the computational complexity [13].Multiple paths can be formed for both traffic sources and intermediate nodes with new routes being discovered only when needed, reducing route discovery latency and routing overheads. Multiple paths can also balance network load by forwarding data packets on multiple paths at the same time [14].

2.1. Selection of OLSR protocol for modification

Optimized Link State Routing (OLSR) protocol selection is done from the view point of implementing multipath technique efficiently due to proactive nature of protocol. In proactive type of protocol energy management is the key issue.

The main concept used in OLSR is that of multipoint relays (MPRs). MPRs are the selected nodes from one hop neighborhood which forward broadcast messages during the flooding process. This technique substantially reduces the message overhead as compared to a classical flooding mechanism (where every node retransmits each message received). In this way, nodes learn their local vicinity and the status of the link with each neighbor. The MPR set for a given node consists of a subset of neighboring nodes that covers the whole two-hop neighborhood of this node. Nodes learn their two-hop neighbor set by exchanging periodic HELLO messages.

This information is disseminated throughout the whole network via periodic Topology Control (TC) messages. This allows mobile nodes to set up routes to any potential destination present in the network. The TC message for a given node contains the set of nodes that have selected the sending node as an MPR. Once a node receives TC messages from other nodes, it can create routing directions to every node in the network using some sort of shortest path algorithm. This way a mobile host can reduce battery consumption.

OLSR provides optimal routes in terms of number of hops. The protocol is particularly suitable for large and dense networks as the technique of MPRs works well in this context. As a proactive protocol, OLSR is also suitable for scenarios where the communicating pairs change over time. No additional control traffic is generated in this situation since routes are maintained for all known destinations at all times [15]. The optimization achieved using the MPRs works well for large and dense mobile networks. The routing table is updated when a change is detected in either: the link set, the neighbor set, the 2-hop neighbor set, the topology set, or the Multiple Interface Association Information Base. Compared to On-Demand routing, proactive routing broadcasts more control messages in the network. If designed properly, these control messages could be effectively utilized to update the route table with only small overhead using multipath approach. In case of multi path OLSR, the routing table stores at the most two routes to every destination in the network. If the major route collapses, the other alternate route can be used immediately without another route discovery, thus, providing better Quality of Service (QoS) than the single route OLSR [16].

3. MODIFICATION SCHEME USED TO MAKE OLSR ENERGY EFFICIENT

We propose a modified protocol, including multipath and energy aware technique in OLSR. The modified protocol can be regarded as a hybrid multipath routing protocol. By using combined approach (multipath and energy aware technique), we expect a more fair distribution of the load along with even utilization of energy resources in the network so as to increase network lifetime as well as individual node lifetime in various dynamic conditions.

In order to include modifications in new protocol we have used reserved field available in the HELLO and TC packet format in original OLSR protocol, keeping all the remaining packets formats same. Reserved bits are modified by lifetime information in our modified protocol. Multipath source routing approach is used in association with the Min-Max Lifetime (MML) as an improvement over the conventional hop by hop routing in original OLSR protocol.

For the protocol modification, following changes are carried out at various stages:

3.1. Calculation of willingness of MPR nodes

An energy-aware selection of willingness can introduce an improvement in MPR selection, allowing the nodes to declare a willingness value of WILL_HIGH (meaning a high willingness to act as a MPR for its neighbors) or WILL_LOW (to signal a low willingness to forward neighbor's data). This way, a node can change its probability to be selected by its neighbors as a MPR, according to its own energy status [17]. In the default implementation of OLSR protocol, every node declares to its neighbors the same willingness (a value named WILL_DEFAULT).

Thus in OLSR, each node has the same probability to be selected as a MPR by its neighbors, and the selection is performed only on the basis of the position of nodes.

In our protocol, the MPR selection is done when node energy is highest (W_High). If such node is not available, then node with medium energy (W_ Default) is selected. Node with less energy (W_Low) is never selected, even when it is the nearest node. It constructs an MPR-set that enables a node to reach any node in the symmetrical strict 2-hop neighborhood through relaying by one MPR node with willingness different from WILL_NEVER or WILL_LOW.

3.2. Topology Sensing

Due to correct MPR selection the flooding of broadcast packets in the network is avoided by reducing duplicate retransmissions in the same region. Multiple possible paths between source and destination pair are maintained instead of a single path. In order to take advantage of multiple paths we have increased TC packet frequency. This explores all possible paths to a particular destination. For our simulation, we have taken TOP_HOLD_TIME as 6 sec. TOP_HOLD_TIME [18] is given by Equation (1). For keeping track of movement of nodes we have reduced the TOP_HOLD_TIME, so that latest updates can be made available with less delay.

TOP_HOLD_TIME = 3*TC_INTERVAL   --- Equation (1)

3.3. Routes Computation

The computation of routes uses the Multipath Algorithm to populate the multipath based on the information obtained from the topology sensing. In our modified protocol, routing table is calculated using Lifetime metric from the information maintained in neighbor set (accumulated from HELLO) & topology set (accumulated from TC). Life time is predicted as ratio of residual battery energy to drain rate which is function of network traffic. The considerations for drain rate are according to current traffic conditions along with residual battery power. It provides a more optimized solution by considering the link traffic in an active network. The method used by each node to calculate the drain rate is similar to running average. Cost function is defined as the ratio of current remaining energy level to drain rate. The cost function is inversely proportional to the network resources used, if the data transmission is to be carried out by that node to its neighbors. This function is then added to the TC as well as to the HELLO packet. The destination node now selects the path in which the least cost function is highest among a set of routes in routing table, leading to destination. The route request packet consists of an IP header.

3.4. Source Routing in modified protocol

In OLSR, due to next hop routing, node can forward data based on its own routing table. It may not get the correct next node from the nodes available to reach the destination. To avoid the problem for the next hop routing in standard OLSR protocol, we use the source path in our modified algorithm. In the routing table of our protocol, information related to hops from destination to source is stored. Source routing will help the source node to keep good control of the packets which will be forwarded in the multipath.

3.5. Route Recovery

In the classical OLSR, the hop-by-hop routing is used, which means when a packet reaches an intermediate node, the protocol will check the routing table of the local node and then forward the packet to the next hop. However, in the mean time, the pure source routing might cause two problems:

- Firstly, the information in the source node might not be the latest because it needs time to flood the topology control messages to the whole network, i.e. while computing the routes; the source node might use the links that do not exist anymore.
- Secondly, even when the information in the source node is updated, the topology might change during the forwarding of the packet.

Both situations will cause the failure of the packet forwarding.

To overcome this problem, we carry out route recovery as explained in step 6 of Implementation for modified Algorithm. Functional representation of the modified protocol is given in Figure 1.

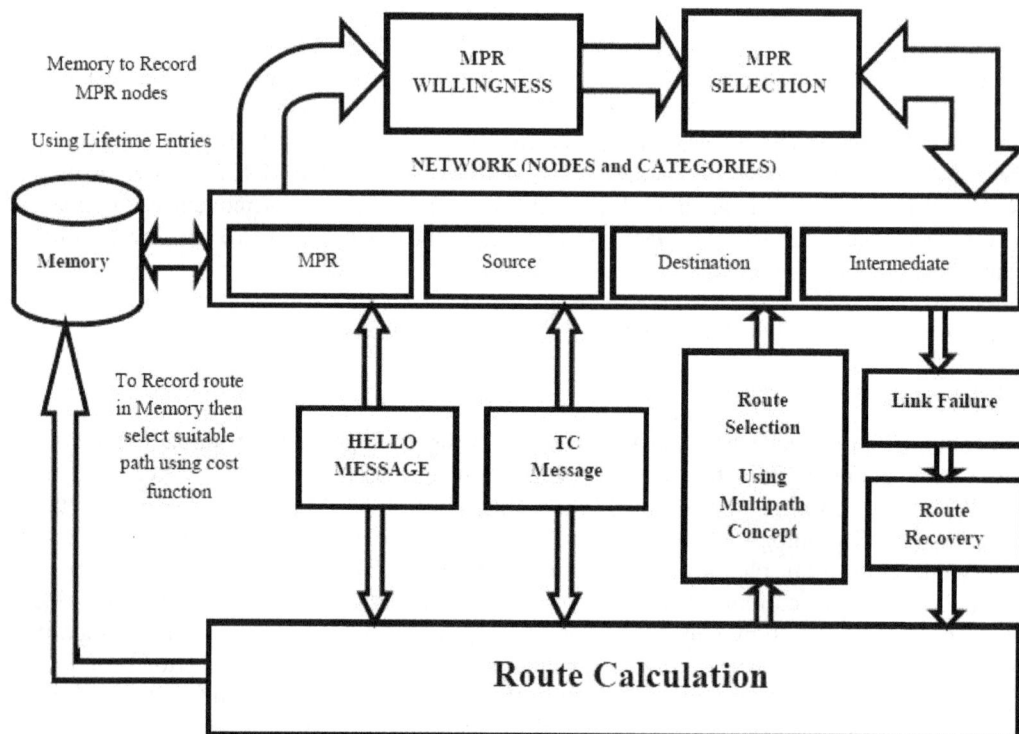

Figure 1. Functional representation of the modified protocol

3.6. Implementation steps for making OLSR energy efficient

1. Remove all previous entries from MPR table.
2. Selection of MPR based on lifetime calculation showing impact of willingness.
3. Based on its neighbor table; insert the new entry to its routing table along with lifetime information.
4. Insert the new entry to routing table along with lifetime values of nodes on path and record the complete path information in the routing table.
5. Perform Route calculation based on following steps:
    a. Remove all previous entries for route.
    b. Add new routing entries with symmetric neighbours as intermediate single-hop neighbours.
    c. Record the new route entry by incrementing hop count by one each time till the destination node is reached.
    d. Records several route entries for same destination with different cost values i.e. multiple paths are stored in memory for single destination called multipath routing.

e.   Out of these multiple paths, select the one with minimum cost value for delivering message to the destination.

As our network is mobile, each node changes its position periodically so that there may be possibility of link breakages. Hence, the concept of route recovery is introduced to get another route to destination.

6.   Route Recovery: For intermediate node, check if the next hop in the source route of the packet is one of its neighbors, before trying to forward the packet,

-   If yes, then forward the packet as it is.
-   If no, then recompute the route to destination and then forward the packet through the new route (This represents the case when the "next hop" has moved out of the transmission range of the node).

3.7. Performance criteria

We use the metric 'Number of nodes alive' in our simulation trials. This metric indicates the overall lifetime of the network [19]. In wireless Ad Hoc networks, especially in those with densely distributed nodes, the death of the first node seldom leads to the total failure of the network. With number of dead nodes increasing, the network is likely to partition. More importantly, it gives an idea of the area coverage of the network over time. Network lifetime is one of important metrics to evaluate the energy efficiency of the routing protocols with respect to network partition [20].

4. SIMULATION

Simulation is carried out for OLSR and modified protocol. Ns-2 is used to compare ad hoc routing protocols over Random waypoint mobility models. The underlying MAC layer protocol is defined by IEEE 802.11 standard [21].

We have measured 'Number of nodes alive' for both the protocols. We observed the effect of variation of pause time, speed and number of nodes on both the protocols. As we are considering node mobility issue, both, speed and pause time play major role. Depending upon their values, the nodes move making the scenario dynamic, leading to few path breaks. Thus the route recovery feature gets tested. Node density shows the network state, sparse or dense, and whether it affects protocol behavior.

Simulations are carried out for three input parameter variations. They are listed at the top of three columns in Table I. The variable parameter entry in a column is varied only for that particular parameter. Other parameter values are same for the three cases.

We have considered field size as 670*670 sq. meters by referring most of simulation trials to earlier work. In order to have sufficient observation interval, simulation time is selected as 1000s. From our previous trial experience we have selected total nodes present in the network as 35 [22]. Number of connections among the nodes is usually half the node count and accordingly the value is selected. A medium data packet rate is selected for the connections. Initial node energy and communication range are interrelated. If node energy is more, then communication range can be large and vice versa. We have selected moderate values of these parameters.

4.1 Considerations for mobility model used for simulation

The performance of ad hoc routing protocols greatly depends on the mobility model used. Random Waypoint is considered to be an entirely random scheme and intuitively can be the most challenging environment for ad hoc routing protocols [23]. Most of the multipath protocols & the energy aware protocols are analyzed in static environment. We analyze our protocol using random way point mobility model which provides worst case test conditions with various dynamic situations.

5. SIMULATION RESULTS FOR MODIFIED PROTOCOL AND OLSR

We observed the effect of number of nodes, speed and pause time variation on both the protocols' simulated parameters. Number of nodes i.e. Node density for fixed area shows the network state (sparse or dense) and how it affects protocol behavior. Number of connections among the nodes should be half that of number of nodes, and should be changed as per nodes under consideration, for the trial. Network mobility effect can be studied by varying speed and pause time. For the trials, we have varied one parameter at a time, keeping other parameters constant. Simulations scenario is given in Table 1.

Table 1. Parameter selection used for the simulation in Ns-2

Parameters	Pause time (seconds)	Speed (meter/sec.)	Number of nodes
Number of mobile nodes	35	35	10,20,30,40,50,60
Field size (m)	670*670	670*670	670*670
Simulation duration (s)	1000	1000	1000
Number of connections	17	17	5,10,15,20,25,30
Speed (m/s)	5	1,5,10,15,20,25,30	5
Pause time (s)	50, 100, 150, 200, 250, 300, 350, 400	100	100
Mac Layer	IEEE 802.11	IEEE 802.11	IEEE 802.11
Communication range (m)	100	100	100
Data packet rate for each connection (packets/s)	20	20	20
Initial node energy (J)	10	10	10
Mobility Model	Random Waypoint	Random Waypoint	Random Waypoint

The comparative analysis of both the protocols for energy related parameter as Number of nodes alive is given in the following sections.

5.1. Effect of pause time

In simulation, pause time can vary from minimum zero sec to half of simulation time. For our simulation purpose we have varied the pause time from 50 to 400 sec. The impact of pause time on Number of nodes alive is tested.

The comparison of Number of nodes alive versus pause time variation for both the protocols is given in Table 2. It is observed that our protocol shows small improvement in Number of nodes alive as compared to OLSR for the pause time values ranging between 150 sec to 350 sec.

Table 2. Effect of pause time variation on Number of nodes alive.

Pause Time (sec)	Number of nodes alive - OLSR	Number of nodes alive - Modified protocol
50	0	0
100	1	1
150	4	7
200	10	11
250	11	13
300	13	15
350	14	16
400	23	23

5.2. Effect of speed

During simulation, we have considered speed variation from 1 m/s to 30 m/s in steps of 5 m/s, to demonstrate different node movement cases. The lowest speed represents person walking, speed as 1 m/s i.e. 3.6 km/hr and other case is that of a moving vehicle with speed 30 m/s or 108 km/hr for fast mobility applications. The considerations are based on Indian scenario.

The Number of nodes alive condition is better maintained for the modified protocol. It is observed that at medium speed i.e. at 15 m/s modified protocol performs better as compared to OLSR as seen in Table 3 .

Table 3. Effect of speed variation on Number of nodes alive

Speed (meter/sec)	Number of nodes alive - OLSR	Number of nodes alive - Modified protocol
1	32	32
5	19	20
10	20	21
15	15	22
20	17	18
25	15	15
30	8	13

5.3. Effect of number of nodes

When topography under consideration is kept constant and we vary number of nodes from 10 to 60 in steps of 10 nodes, the effect of nodes variation on Number of nodes alive is observed. It is observed from Table 4, that, modified protocol shows improvement in Number of nodes alive as compared to OLSR when network contains more number of nodes i.e. 60 nodes. This confirms the effective use of our protocol for dense networks.

Table 4. Effect of number of nodes variation on Number of nodes alive

Number of Nodes	Number of nodes alive - OLSR	Number of nodes alive - Modified protocol
10	6	6
20	15	16
30	17	17
40	18	18
50	19	20
60	20	25

The protocol testing is further carried out for Simulation time variation for two sample values of pause time, speed and number of nodes. These trials help to decide better choice of these three parameters for network performance.

5.4. Simulation time variation for two cases of pause time

Two sample values of Pause time (250 and 400 sec) are selected from the range 50 to 400 sec for simulation time variation trials. We have carried out simulations for twice the selected pause time value. The comparison of protocols for both cases, for same simulation time duration is given in Figure 2. The results show modified protocol is superior as compared to OLSR for both pause time conditions, from 300 sec simulation time. Number of nodes alive is more for 400 sec as compared to 250 sec.

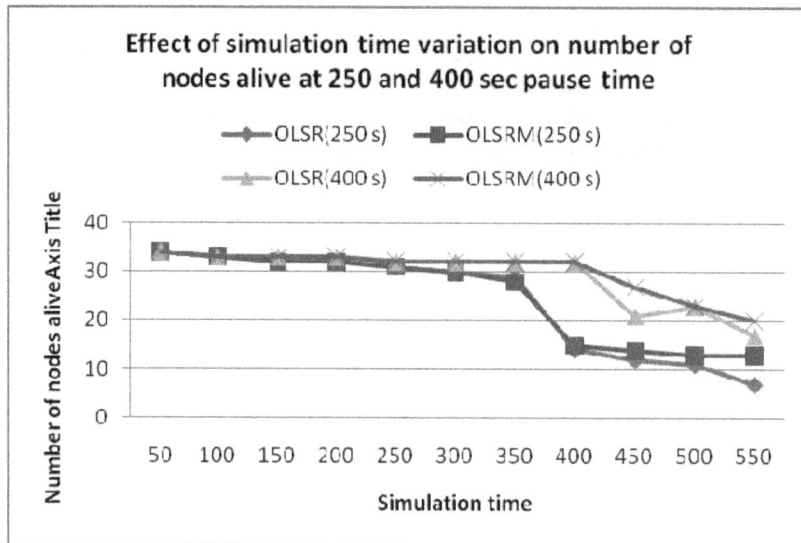

Figure 2. Effect of simulation time variation on number of nodes alive, for OLSR and modified protocol alive at 250 sec and 400 sec pause time

5.5. Simulation time variation for two cases of speed

Number of nodes alive against simulation time at 1m/s and 15 m/s speed for OLSR and modified protocol is given in Figure 3. No significant difference is observed in number of node alive, for both protocols except for simulation time as 450 and 500 sec for both speed conditions.

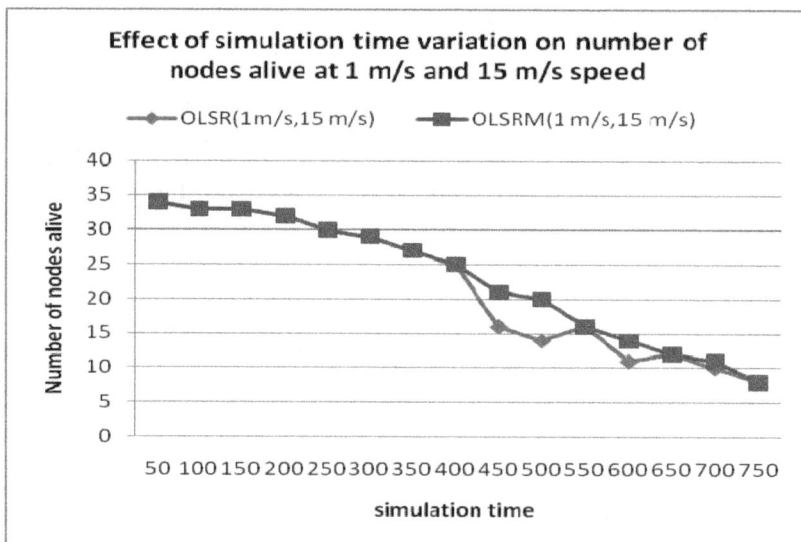

Figure 3. Effect of simulation time variation on number of nodes alive, for OLSR and modified protocol at 1 m/s and 15 m/s speed conditions

5.6. Simulation time variation for two cases of number of nodes

Number of nodes alive against simulation time with 40 and 60 nodes for OLSR and modified protocol is given in Figure 4. When number of nodes selected for given scenario is more, initially when simulation starts, the node count decreases earlier. Later, over total simulation time trials, Number of nodes alive for modified protocol is better.

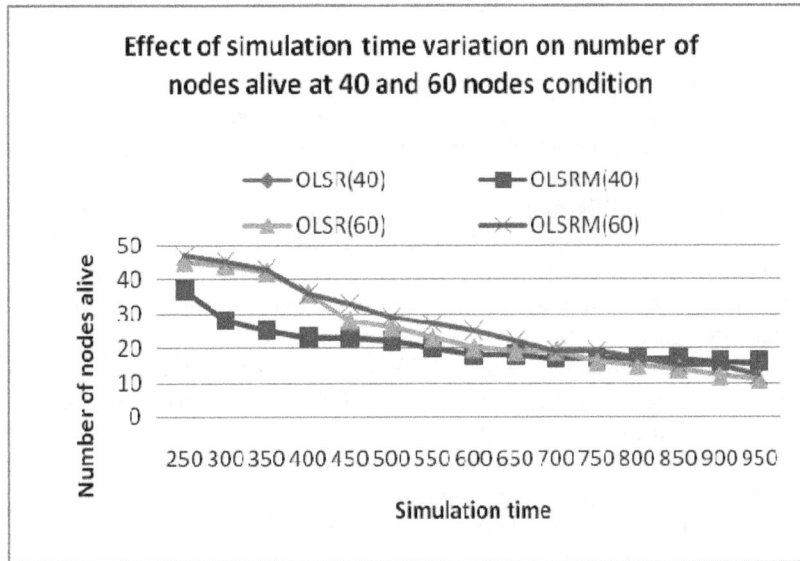

Figure 4. Effect of simulation time variation on number of nodes alive, for OLSR and modified protocol at 40 and 60 number of nodes conditions

6. PETRI NET

The main benefit of modeling is that it provides insight about the properties of the system prior to implementation. This allows many issues about the system to be resolved in the design phase rather than in the implementation phase.

The use of formal description techniques results in models that are amenable to verification. An advantage of many formal description techniques is that, they are based on the construction of executable models that make it possible to observe and experiment with the behavior of the protocol prior to implementation using simulation. This typically leads to complete specifications since the model will not be fully operational until all parts of the protocol have been specified. Another advantage of formal modeling languages is that they support abstractions, making it possible to specify the operation of the protocol without being concerned with irrelevant implementation details. The complex behavior of communication protocol makes the design of correct protocols a challenging task due to number of independent concurrent protocol entities.

Petri nets are a fundamental, visual and formal modeling technique in concurrency and have been subject for suitable extensions in particular to model reconfigurations of the net structure and the exchange of data. Thus they are profitably applied to model workflows in mobile ad-hoc networks and flexible communication based systems [24]. Petri net (PN) have an appealing conceptual simplicity based on graphical representation of the mechanism of process interaction. Due to the dynamic nature, concurrency and different levels of abstraction associated with the Mobile Ad-Hoc Network (MANET) protocols, Colored Petri Nets (CPN) is a suitable modeling language for this purpose. This is a promising tool for describing and studying information processing systems that are characterized as being concurrent, asynchronous, distributed, parallel, nondeterministic and stochastic [25]. Petri net being state transition based model can be used widely for protocol studies for handling concurrent or communication systems. PN provides only a structural description of protocols; the dynamic aspects of protocol like control and data flow are described in terms of firing rules and token distribution [26]. High level Petri nets are powerful for reliable specification for translation of requirement definitions into verified specifications. High level Petri net goes through various phases such as requirement definition, validation, specification and implementation [27].

CP-nets or CPNs is a formal modeling language that is well suited for modeling and analyzing large and complex systems for several reasons: hierarchical models can be constructed, complex information can be represented in the token values and inscriptions of the models, timing information can be included in the models, and mature and well-tested tools exist for creating, simulating, and analyzing CPN models [28]. CP-nets are often used to model and verify the logical correctness of network protocols. Colored Petri nets provide a framework for the design, specification, validation, and verification of systems. CP-nets have a wide range of application areas like communication protocols, operating systems, hardware designs, embedded systems, software system designs, and business process re-engineering. Design/CPN is a graphical computer tool supporting the practical use of CP-nets [29]. The tool supports the construction, simulation, functional and performance analysis of CPN models. CPN models can be structured into a number of related modules. The module concept of CP-nets is based on a hierarchical structuring mechanism (either top down or bottom up approach). CP-nets include a time concept which makes it possible to capture the time taken by different activities in the system. Timed CPN models and simulation can be used to analyze the performance of a system, by investigating Quality of Service (QoS) parameters like, delay, and throughput. It is possible to investigate the functional correctness of systems modeled by means of timed CP-nets. Abstract CPN models can be used in an early phase of system development to determine the boundaries of the project and specify requirements [30]. CP-nets have a sound, mathematically well founded execution semantics, are well-proven, and have proper tool support. The design and specification can be supported by modeling and simulation using Hierarchical Colored Petri Nets (CP–nets). The main purpose of these models has been to analyze the behavior of existing systems, it can be considered as an integrated part of the design phases (which is more time consuming than actual coding) of the development of distributed software systems. The use of CP-nets in the design phase contributed to the development of a better product using fewer resources [31]. Hierarchical nets provide construction of complicated models [32]. In such nets an element may be represented by another net i. e. nested construction: net inside net. The number of hierarchy levels has no principal limitations like programming languages where procedures are used to maintain the complexity.

The CPN modeling language and supporting computer tools are powerful enough to specify and analyze a real-world communication protocol [33]. The act of constructing the CPN model, executing, and discussing it lead to the identification of several non-trivial design errors and issues that under normal circumstances would not have been discovered until at best in the implementation phase.

6.1. Verification of modified energy efficient OLSR protocol using Hierarchical Colored Petri Nets (HCPN)

Main-body As protocol modification is done at different parts, we have used HCPN technique. Figures 6 to 8 illustrate separate mechanisms that are used to build modified OLSR as given in Figure 5. For practical reasons it is not desirable to create a single large CPN that specifies a given firewall system in a flat structure. The concept of Hierarchical CPNs allows a designer to construct large CPNs by combining a number of smaller CPNs. They are beneficial for the modular composition of CPNs. HCPNs can be constructed top-down, bottom-up, or by a mixture of these two strategies. We are using top down approach for HCPN. HCPNs make it possible to relate a number of individual CPNs to each other in a formal way, and thus allow their formal analysis. In a top-down design one starts with a simple high level description of a system without consideration for internal details. A specification of detailed behavior of the CPN is developed through stepwise refinement. Stepwise refinement is achieved through the application of a construct called substitution transition, where a more complex CPN takes the place of a transition. The CPN must conform to the interface of the replaced transition and relate identically to its surrounding arcs. In a bottom-up design CPNs are combined into a larger net through fusion places. A fusion place is a set of places that are considered to be identical. Even if they are drawn as individual places they represent a single conceptual place. For each token that is added (removed) at one of the places, an identical token is added (removed) at all other places. Point 'A' in Figure 6 and in Figure 7 is a fusion place, as it appears in both places identically. A non-hierarchical CPN is called a page. A page that contains a substitution transition is called a superpage as shown in Figure 6. A page that contains the detailed description of the activity modeled by the corresponding substitution transition is called subpage given in Figure 5. A substitution transition is also called a supernode. Note that the places connected to a substitution transition by a single arc (called socket nodes) and their counterparts on the subpage (called port nodes) are fusion places. The interface between a superpage and a subpage is defined through port assignments where socket nodes are related to port nodes.

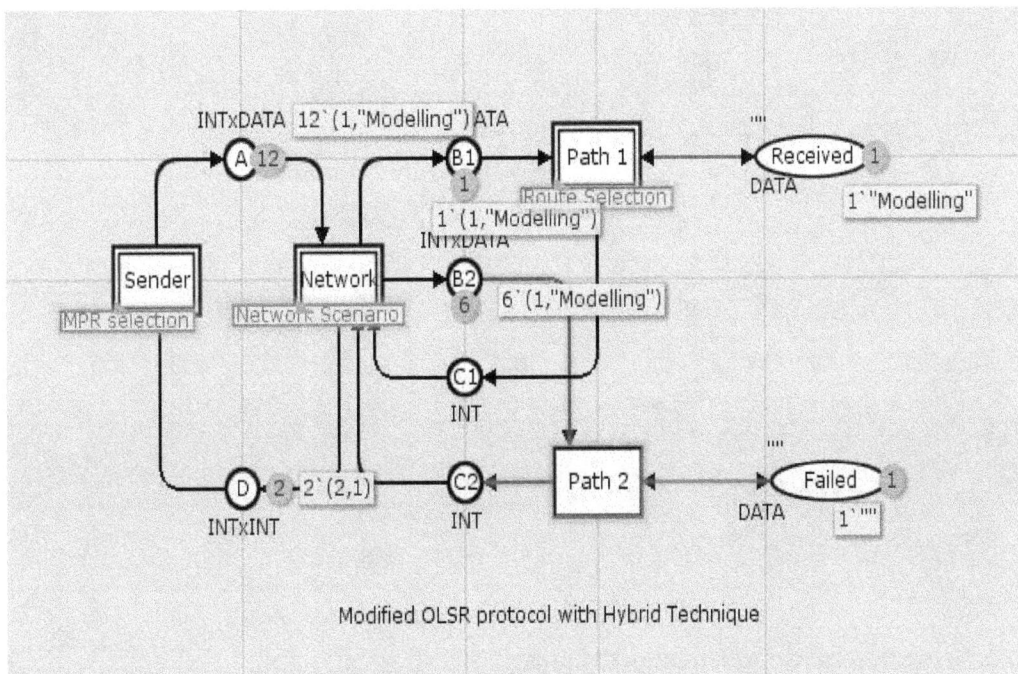

Figure 5. Modified protocol with Hybrid Technique represented in CPN

6.2. Working

Figure 5 shows whole functioning of Modified OLSR protocol, which is further divided into three parts.

6.2.1   MPR Selection criteria (Figure 6)

6.2.2   Network Scenario with multipath selection (Figure 7)

6.2.3   Energy Efficient Route Selection (Figure 8)

HCPN represented by Figure 5, describes an example of Modified OLSR that combines MPR selection and Route Selection mechanisms. We structure the description top-down, starting with the superpage. Sender sends a packet to the Network by creating a copy of the packet on place A. It should be noted that neither the packet is removed from Send nor the counter at Next Send increased. The reason for retaining the packet is that, the packet may be lost and hence should be retransmitted. Our protocol continues to repeat the same packet until it gets an acknowledgement telling that the packet has been successfully received. Transition shows a packet transmits from the Sender site of the Network to the Receiver site by comparing cost value for different paths and moving the corresponding token from A to B. It should be noted that the Route Selection page is used by two substitution transitions, Received and Failed. This means that we will have two instances of the subnet – during a simulation. The two instances may have different markings and different enabling depending on route selection.

6.2.1. MPR Selection

The MPR selection part is used to select node as MPR depending on WILLINGNESS. The only difference is that Receive Acknowledgment now needs an acknowledgment from Receiver in order to become enabled.

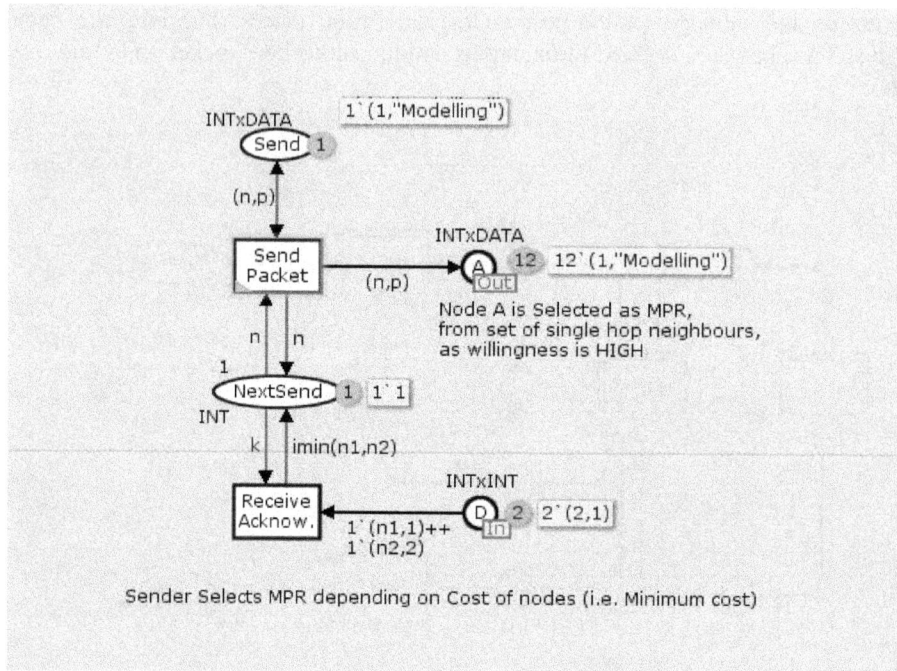

Figure 6. MPR Selection Criteria based on node cost

6.2.2. Network Scenario with multipath option

Transmit Packet produces packets at two different output places B1 and B2 (i.e. nodes). The packets at B1and B2 are for the Receiver. It should be noted that we use two different variables r1 and r2 to determine whether the packets for B1 and B2 are lost or not. This means that we model a broadcast in which one of the paths may get a packet while the other does not. Transmit Acknowledgment is split in two parts in case of path failure. The Transmit acknowledgment can be from C1 on the path (B1) or from C2 on the path (B2). B1 and B2 modify the acknowledgment, by adding information telling the Sender from where the acknowledgment came. The Sender selects energy efficient path among B1 and B2. The Boolean expression Ok (s, r) determines whether the packet is successfully transmitted or lost. Each acknowledgment is a pair (C1, B1) or (C2, B2) where the first element is contents, while the second element indicates Path as shown in Figure 7.

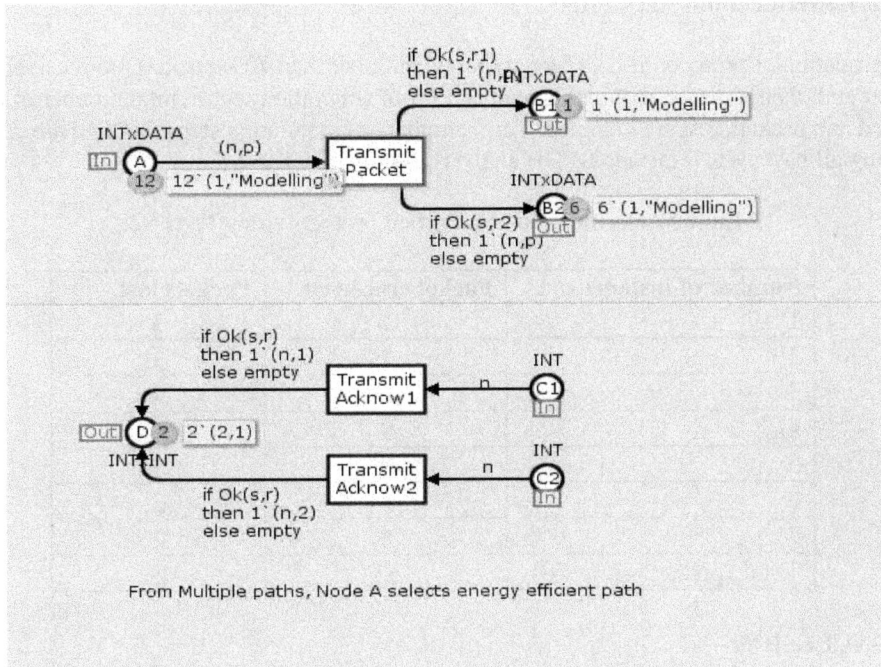

Figure 7. Network Scenario with Multipath option

6.2.3. Energy Efficient Route Selection

Path having minimum cost is selected for transmission. The desirable path B1 is denoted by B; similarly contents C1 are denoted as C. This is route selection criteria shown in Figure 8.

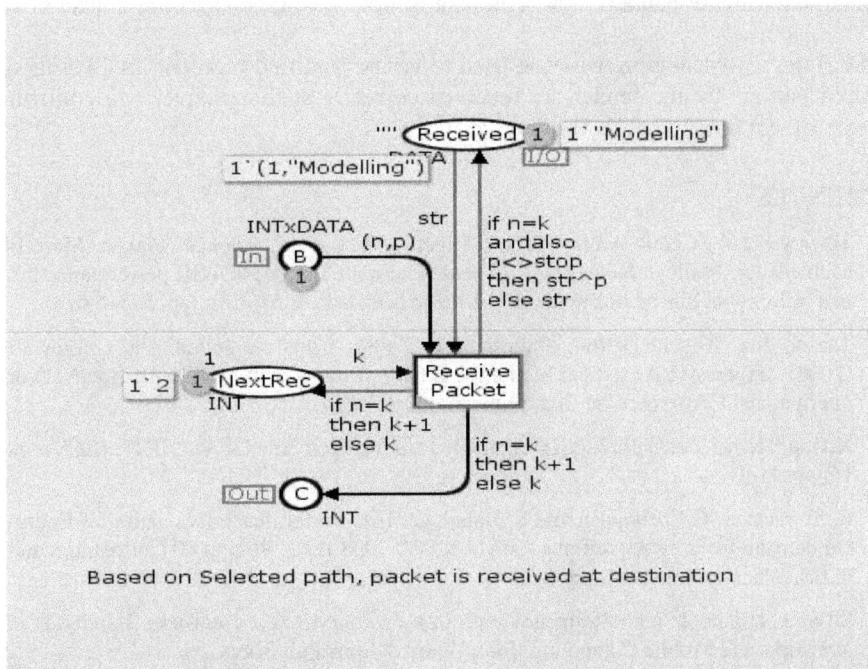

Figure 8. Energy Efficient Route Selection

6.3. OLSR verification:

For verification of protocol, it is expected that when more than 70% packets are received at receiver end, then it is said protocol is realistic. From simulation report, total number of packets received and packets lost are obtained. Total numbers of simulation steps required are 205 for receiving all packets at 7 instances. The analysis is shown in Table 5.

Table 5. Verification of modified protocol using Petri Net

Number of instances	Packets received	Packets lost
1	7	3
2	7	3
3	6	4
4	9	1
5	8	2
6	8	2
7	8	2
Total (packets sent 70)	53	17

7. CONCLUSION

Our modified protocol, achieves better energy efficiency as compared to OLSR in terms of Number of node alive, indicating better network lifetime. For modified protocol, the improvement in Number of nodes alive for pause time and speed variation is due to multipath technique included in the protocol in addition to cost function and source routing. Modified protocol shows more energy efficient behavior for denser network. This is because in denser network there is better chance of balancing the traffic through many different energy efficient paths. Simulation time variation trials show that for better value of Number of nodes alive, it is desirable to have higher pause time and number of nodes, while no specific impact of speed.

In our Petri net representation, we have tried to verify modified protocol. In CPN model, as the transmitted packets by the sender are received correctly at the receiver, we confirm that, the modified protocol functioning is verified.

REFERENCES

[1]   Thierry Plesse, Cedric Adjih, Pascale Minet, Anis Laouiti, Adokoe´ Plakoo, Marc Badel, Paul Muhlethaler, Philippe Jacquet and Je´rome Lecomte, (2005) "OLSR performance measurement in a military mobile ad hoc network", *Ad Hoc Networks 3, Elsevier,* pp. 575–588.

[2]   Taesoo Jun, Angela Dalton, Shreeshankar Bodas, Christine Julien, and Sriram Vishwanath, (2008) "Expressive Analytical Model for Routing Protocols in Mobile Ad Hoc Networks", IEEE international Conference on Communication, pp. 1-7.

[3]   X.Hong, K.Xu and Gerla,(2002) "Scalable Routing Protocols for MANET", *IEEE network,* Vol. 16, pp. 11-21.

[4]   P. Sivasankar, C.Chellappan and S. Balaji, (2010) "Performance Evaluation of Energy Efficient On demand Routing Algorithms for MANET", 2008 IEEE Region 10 Colloquium and the Third ICIIS, Kharagpur, INDIA, pp. 1-5.

[5]   Silvia Giordano & Ivan Stojmenovic (2002) *Mobile Ad Hoc Networks,* Handbook of Wireless Networks and Mobile Computing, John Wiley & Sons Publishers.

[6]   Jiazi Yi, Eddy Cizeron, Salima Hamma and Benoît Parrein, (2008) "Simulation and Performance Analysis of MP-OLSR for Mobile Ad hoc Networks", IEEE WCNC, pp. 2235-2240.

[7]   F. Chiti, M. Ciabatti, G. Collodi, D. Di Palma, R. Fantacci, and A. Manes, (2006), "Design and application of enhanced communication protocols for wireless sensor networks operating in environmental monitoring", Proceedings of IEEE International Conference on Communications (ICC '06), Istanbul, Turkey, Vol. 8, pp. 3390–3395.

[8]   Geetha Jayakumar, and G. Gopinath, (2007) "Ad Hoc Mobile Wireless Networks Routing Protocols – A Review", *Journal of Computer Science*, 3 (8), pp. 574-582.

[9]   Ajit Singh, Harshit Tiwari, Alok Vajpayee and Shiva Prakash, (2010) " A Survey of Energy Efficient Routing Protocols for Mobile Ad-hoc Networks", *International Journal on Computer Science and Engineering (IJCSE),* Vol. 02, No. 09, pp. 3111-3119.

[10] Saoucene Mahfoudh and Pascale Minet, (2008) " An energy efficient routing based on OLSR in wireless ad hoc and sensor networks", 22nd International Conference on Advanced Information Networking and Applications – Workshops IEEE Computer Society, pp. 1253-1259.

[11] C.-K. Toh, (2001) " Maximum Battery Life Routing to Support Ubiquitous Mobile Computing in Wireless Ad Hoc Networks", IEEE Communication Magazine, pp. 138-147.

[12] Stephen Mueller, Rose P. Tsang and Dipack Ghosal, (2004) "Multipath Routing in Mobile Ad Hoc Networks: Issues and Challenges", Lecture Notes in Computer Science, Vol. 2965, pp. 209 – 234.

[13] Mustafa K. Gurcan, Hadhrami Ab Ghani, Jihai Zhou and Anusorn Chungtragarn, (2011) "Bit Energy Consumption Minimization for Multi-path Routing in Ad Hoc Networks", *Computer Journal* , Volume 54, Issue 6, pp. 944-959.

[14] S. R. Biradar, Koushik Majumder, Subir Kumar Sarkar and Puttamadappa, (2010) "Performance Evaluation and Comparison of AODV and AOMDV", *International Journal on Computer Science and Engineering (IJCSE),* Vol. 02, No. 02, pp. 373-377.

[15] Thomas Clausen and Philippe Jacquet,(2010) " The optimized Routing Protocol for Mobile Ad Hoc Networks: Protocol Specification", *INRIA*, version 1, pp. 1-53.

[16] Sun Xuekang, Gu Wanyi, Xiao Xingquan, Xu Baocheng and Guo Zhigang, (2009) " Node Discovery Algorithm Based Multipath OLSR Routing Protoco"l, WASE International Conference on Information Engineering, IEEE Computer Society, pp. 139-142.

[17] Floriano De Rango and Marco Fotino, (2009) "Energy Efficient OLSR Performance Evaluation under Energy aware Metrics", SPECTS, pp. 193-198.

[18] T. Clausen and P. Jacquet, (2003), "Optimized Link State Routing Protocol (OLSR)", RFC-3626, pp. 1-75.

[19] Arati Manjeshwar and Dharma P. Agrawal, (2001) "TEEN: A Routing Protocol for Enhanced Efficiency in Wireless Sensor Networks", Fifteenth International Parallel and Distributed Processing Symposium (IPDPS'01) Workshops, Vol. 3, pp. 30189a.

[20] Alemneh Adane, (2008) *Active Communication Energy Efficient Routing Protocol of Mobile Ad Hoc Networks (MANETS)*, PhD Thesis, Addis Ababa University, Computer Engineering, Ethiopia.

[21] IEEE Computer Society LAN MAN Standards Committee,(1999) *Wireless LAN Medium Access Control and Physical Layer Specifications*, in IEEE 802.11 Standard.

[22] Radhika D. Joshi and Priti P. Rege, (2009) "Application Specific Node Selection Criteria for Mobile Ad Hoc Networks", *CIIT International Journal of Wireless Communication*, CIIT-IJ-0386, Vol. 1, No. 6, pp. 274-283.

[23] Bor-rong Chen and C. Hwa Chang, (2003) " Mobility Impact on Energy Conservation of Ad Hoc Routing Protocols", Proceedings of International Conference on Advances in Infrastructure for Electronics, Business, Education, Science, Medicine and Mobile Technologies on the Internet (SSGRR'03) MTC, pp. 1-7.

[24]   Kathrin Hoffmann, (2010) "Formal Modeling and Analysis of Mobile AdHoc Networks and Communication Based Systems using Graph and Net Technologies", Bulletin of the European Association for Theoretical Computer Science (EATCS), No.101, pp.148-160.

[25]   Amirineni, K K Chinara, S Rath, S K , (2010), "Validation of Clustering Algorithm for Mobile Ad-Hoc Networks Using Colored Petri Nets", Proceedings of the 4th National Conference; Bharati Vidyapeeth's Institute of Computer Applications and Management, New Delhi INDIACom, Computing For Nation Development.

[26]   To-yat Chung, (1996) "Petri Nets for Protocol Engineering, *Elsevier*, Computer communications", pp. 1250-1257.

[27]   Bruno Blaskovic, (1998) "Petri Net Modeling for Signaling Protocol Synthesis", Electro technical conference, Ninth Mediterranean, Melecon, pp. 706 – 710.

[28]   Lisa Wells, (2006) "Performance Analysis using CPN Tools", *ACM*, Value Tools '06, Pisa, Italy, pp. 1-10.

[29]   Lars M. Kristensen, Soren Christensen and Kurt Jensen, (1998) "The practitioner's guide to coloured Petri nets", *Springer Verlag, International Journal STTT*, 2, pp. 98-132.

[30]   Lars Michael Kristensen, Jens Bæk Jorgensen, and Kurt Jensen, (2004) "Application of Coloured Petri Nets in System Development", *Springer-Verlag, ACPN 2003, LNCS 3098*, pp. 626–685.

[31]   Soren Christensen and Leif Obel Jepsen, (1991) "Modeling and Simulation of a Network Management System using Hierarchical Coloured Petri Nets", In Erik Mosekilde, Proceedings of the Simulation Multiconference, pp. 1-11.

[32]   D.A. Zaitsev and T.R. Shmeleva, (2006) *Simulating of Telecommunication Systems with CPN Tools*, Students' book on the course - Mathematical Modeling of Information Systems, Transaction No. 5.

[33]   Lars Michael Kristensen and Kurt Jensen, (2004) "Specification and Validation of an Edge Router Discovery Protocol for Mobile Ad Hoc Networks", *Springer-Verlag, INT 2004, LNCS 3147*, pp. 248–269.

9

Weighted Dynamic Distributed Clustering Protocol for Heterogeneous Wireless Sensor Network

Said Benkirane[1], Abderrahim Beni hssane[1], Moulay Lahcen Hasnaoui[2], Mostafa Saadi[1] and Mohamed Laghdir[1]

[1]MATIC laboratory, Department of Mathematics and Computer Science, Chouaïb Doukkali University, Faculty of Sciences El Jadida, Morocco
`sabenk1@hotmail.com, abenihssane@yahoo.fr, saadi_mo@yahoo.fr, laghdirm@yahoo.fr`
[2]Computer Science Department, Faculty of Sciences Dhar el Mahraz, Sidi Mohammed Ben Abdellah University, Fez, Morocco.
`mlhnet2002@yahoo.ca`

ABSTRACT

In wireless sensor networks (WSN), conserving energy and increasing lifetime of the network are a critical issue that has been addressed by substantial research works. The clustering technique has been proven particularly energy-efficient in WSN. The nodes form groups (clusters) that include one cluster head and member clusters. Cluster heads (CHs) are able to process, filter, gather the data sent by sensors belonging to their cluster and send it to the base station. Many routing protocols which have been proposed are based on heterogeneity and use the clustering scheme such as SEP and DEEC.

In this paper we introduce a new approach called WDDC in which cluster heads are chosen on the basis of probability of ratio of residual energy and average energy of the network. It also takes into consideration distances between nodes and the base station to favor near nodes with more energy to be cluster heads. Furthermore, WDDC is dynamic; it divides network lifetime in two zones in which it changes its behavior.

Simulation results show that our approach performs better than the other distributed clustering protocols such as SEP and DEEC in terms of energy efficiency and lifetime of the network.

KEYWORDS:

Wireless Sensor Networks, Energy Efficiency, Dynamic Protocols, Heterogeneous Network, Clustering.

1. INTRODUCTION

A Wireless Sensor Network (WSN) is composed of hundreds or thousands of sensor nodes. These small sensor nodes contain sensing, data processing and communicating components. A wireless sensor network comprises a base station (BS) that can communicate with a number of wireless sensors via radio link.

The base station in sensor networks is very often a node with high processing power, high storage capacity and the battery used can be rechargeable.

Data are collected at a sensor node and transmitted to the BS directly or by means of other nodes. All collected data for a specific parameter like temperature, pressure, humidity, etc are processed in the BS and then the expected amount of the parameter will be estimated. In these networks, the position of sensor nodes need not be engineered or pre-determined, which allows random deployment in inaccessible terrains or disaster relief operations [1].

Communication protocols highly affect the performance of WSNs by evenly distributing energy load and decreasing their energy consumption and thereupon prolonging their lifetime. Thus, developing energy-efficient protocols is fundamental for prolonging the lifetime of WSNs [2].

Among the proposed communication protocols, hierarchical (cluster based) ones have significant savings in the total energy consumption of wireless micro sensor network [3][4][5]. In these protocols, the sensor nodes are grouped into a set of disjoint clusters. Each cluster has a designated leader, the so-called cluster-head (CH). Nodes in one cluster do not transmit their gathered data directly to the BS, but only to their respective cluster-head.
Besides, it is approved [3] that the use of the clustering technique reduces communication energy more than direct transmission (DT) and minimum transmission-energy (MTE) routing.

In this paper, we propose a new approach, called WDDC, which is based on the ratio of residual energy and average energy of the network and takes into consideration distances between nodes and the base station to determine near nodes and distant nodes in order to give more chance to the nearest nodes to be cluster heads by modifying the election probability value for every type of nodes.

WDDC is dynamic, autonomous and more energy-efficient. Simulation results show that it prolongs the network lifetime much more significantly than the other clustering protocols such as SEP and DEEC.

The remainder of the paper is organized as follows: Section 2 contains the related work done. Section 3 explains the heterogeneous network and radio energy dissipation model. Section 4 describes the DEEC protocol followed by section 5 which describes our WDDC approach. Section 6 shows the simulation results and finally Section 7 gives concluding remarks.

2. RELATED WORK

There are two kinds of clustering schemes. The first kind is called homogeneous clustering protocols. They are applied in homogeneous networks, where all nodes have the same initial energy, such as LEACH [3], PEGASIS [6], and HEED [7]. The second kind of clustering algorithms applied in heterogeneous networks are referred to as heterogeneous clustering schemes [8], where all the nodes of the sensor network are equipped with different amounts of energy, such as SEP [9], M-LEACH [10], EECS [11], LEACH-B [12] and DEEC [13].
WSNs are more likely to be heterogeneous networks than homogeneous ones. Thus, the protocols should be fit for the characteristic of heterogeneous wireless sensor networks. Moreover, in [14, 15], they propose protocols which use a new conception based on the energy left in the network.

Low-Energy Adaptive Clustering Hierarchy (LEACH) [3] is proposed by Heinzelman et al., which is one of the most fundamental protocol frameworks in the literature. LEACH is a clustering-based protocol architecture which utilizes randomized rotation of the Cluster-Heads (CHs) to uniformly distribute the energy budget across the network. The sensor nodes are arranged into several clusters and in each cluster, one of the sensor nodes is chosen to be CH. Each node will transmit its data to its own CH which forwards the sensed data to the BS finally. Both the communication between sensor nodes and CH and that between CHs and the BS are direct, single-hop transmission.

PEGASIS [6] is a chain-based protocol which evades cluster formation and uses only one node in a chain to transmit to the BS instead of using multiple nodes. Heinzelman, et.al. [14] proposed LEACH-centralized (LEACH-C), a protocol that employs a centralized clustering algorithm and the same steady-state protocol as LEACH. O. Younis, et al. [7] proposed HEED (Hybrid Energy-Efficient Distributed clustering), which regularly select cluster heads according to a hybrid of the node residual energy and a secondary parameter, such as node proximity to its neighbors or node degree. G. Smaragdakis, I. Matta and A. Bestavros proposed SEP (Stable Election Protocol) [9] in which every sensor node in a heterogeneous two-level hierarchical network independently elects itself as a cluster head based on its initial energy relative to that of other nodes. Li Qing et.al [13] proposed DEEC (Distributed Energy Efficient Clustering) algorithm in which the cluster head is selected on the basis of probability of ratio of residual energy of each node and the average energy of the network. In this algorithm, a node having more energy has more chances to be a cluster head. This solution doesn't take into account the notion of distances between nodes and the base station, and this protocol is not dynamic.

Among the new researches, we can mention the work done by Latif et al. [16]. In this work, the authors compared four protocols using linear programming formulation technique. It is concluded from their analytical simulation results that DEEC is the most energy-efficient protocol for heterogeneous node energy network.

Another work made by Chang et al. [17] proposes an energy-saving clustering algorithm to provide efficient energy consumption in the network. The approach proposed in this work is to reduce data transmission distance of sensor nodes in wireless sensor networks by using the uniform cluster notion. In order to make an ideal distribution for sensor node clusters, they calculate the average distance between the sensor nodes and take into account the residual energy for selecting the appropriate cluster head nodes. The lifetime of wireless sensor networks is extended by using the uniform cluster location and equalizing the network loading among the clusters.

There is also another work by Rajni et al. [18] in which the authors propose a new clustering protocol for prolonging the network lifetime. The algorithm proposed is the modification of DEEC protocol. It is called a Clustering Technique for Routing in Wireless Sensor Networks (CTRWSN). It is a self-organizing and dynamic clustering method that divides dynamically the network on a number of clusters previously fixed. The operation of CTRWSN is divided into rounds where each round consists of a clustering stage and distributed multi-hop routing stage.

3. HETEROGENEOUS NETWORK AND RADIO ENERGY DISSIPATION MODEL

3.1. Heterogeneous Network Model

In this study, we explain the network model. We suppose that there are N sensor nodes, which are uniformly dispersed within a M x M square region (Figure.1). The nodes always have data to transmit to a base station, which is often far from the sensing area. The network is structured into a clustering hierarchy, and the cluster-heads execute fusion function to reduce correlated data produced by the sensor nodes within the clusters. The cluster-heads transmit the aggregated data directly to the base station. We assume that the nodes are static (not moving). In the two-level heterogeneous networks, there are two types of sensor nodes, i.e., the advanced nodes and normal nodes. Note E_0 the initial energy of the normal nodes, and m the fraction of the advanced nodes, which own a times more energy than the normal ones. Thus there are Nm advanced nodes equipped with initial energy of $E_0(1 + a)$, and $N(1 - m)$ normal nodes equipped with initial energy of E_0. The total initial energy of the two-level heterogeneous networks is given by:

$$E_{total} = N(1 - m)E_0 + NmE_0(1 + a) = NE_0(1 + am) \quad (1)$$

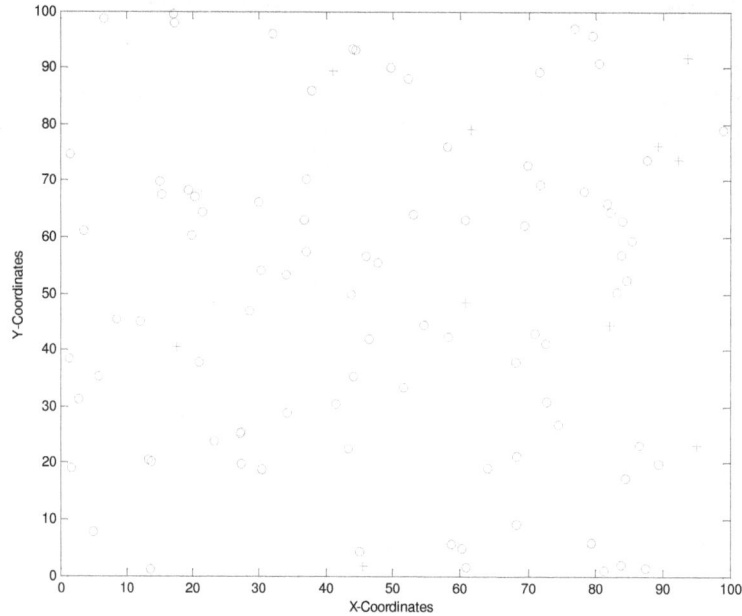

Figure 1. 100 nodes randomly deployed in the network (o normal node, + advanced node).

3.2. Radio Energy Dissipation Model

According to the radio energy dissipation model proposed in [14] (Figure 2) and in order to achieve an acceptable Signal-to-Noise Ratio (SNR) in transmitting an L-bit message over a distance d, the energy expended by the radio is given by:

$$E_{Tx}(l, d) = \begin{cases} lE_{elec} + l\epsilon_{fs}d^2, & d < d_0 \\ lE_{elec} + l\epsilon_{mp}d^4, & d \geq d_0 \end{cases} \quad (2)$$

Where E_{elec} is the energy dissipated per bit to run the transmitter E_{Tx} or the receiver E_{Rx} circuit, ϵ_{fs} is the free space fading energy, ϵ_{mp} is the multi-path fading energy and d is the distance between the sender and the receiver.

To receive this message the radio expends energy:

$$E_{Rx}(l) = lE_{elec} \quad (3)$$

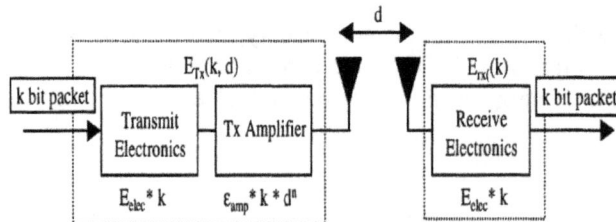

Figure 2. Radio Energy Dissipation Model.

4. DEEC PROTOCOL:

In DEEC [13] the cluster-heads are chosen by a probability based on the ratio between residual energy of each node and the average energy of the network.

As mentioned in section 3.1, the total initial energy of the two-level heterogeneous networks is computed as:

$$E_{total} = N(1-m)E_0 + NmE_0(1+a) = NE_0(1+am) \quad (4)$$

The probability threshold which each node s_i uses to determine whether itself to become a cluster-head in each round is as follows:

$$T(s_i) = \begin{cases} \dfrac{p_i}{1-p_i(r \bmod \frac{1}{p_i})} & if \ \ s_i \in G \\ 0 & otherwise \end{cases} \quad (5)$$

Where G is the group of nodes that are eligible to be cluster heads at round r. In each round r, when node s_i finds itself eligible to be a cluster head, it will choose a random number between 0 and 1. If the number is less than threshold $T(s_i)$, the node s_i becomes a cluster head during the current round.

Also, for two-level heterogeneous networks, p_i is defined as follows:

$$P_i = \begin{cases} \dfrac{p_{opt}E_i(r)}{(1+am)\bar{E}(r)} & \text{if } s_i \text{ is the normal node} \\ \dfrac{p_{opt}(1+a)E_i(r)}{(1+am)\bar{E}(r)} & \text{if } s_i \text{ is the advanced node} \end{cases} \quad (6)$$

Where $E_i(r)$ denotes the residual energy of node s_i and $\bar{E}(r)$ is the average energy of the network at round r.

The estimate value of $\bar{E}(r)$ is:

$$\bar{E}(r) = \frac{1}{N}E_{total}\left(1-\frac{r}{R}\right) \quad (7)$$

Where R indicates the total rounds of the network lifetime.

The value of R can be approximated as:

$$R = \frac{E_{total}}{E_{Round}} \quad (8)$$

Where E_{Round} denote the total energy dissipated in the network during a round r. E_{Round} is given by:

$$E_{Round} = L\left[2NE_{elec} + NE_{DA} + k\epsilon_{mp}d_{toBS}^4 + N\epsilon_{fs}d_{toCH}^2\right] \quad (9)$$

Where k is the number of clusters, E_{DA} is the data aggregation cost expended in CH and BS, d_{toBS} is the average distance between the cluster-head and the base station and d_{toCH} is the average distance between cluster members and the cluster-head.

According to [14][19] we can get the equations as follows:

$$d_{toCH} = \frac{M}{\sqrt{2k\pi}} \ And \ d_{toBS} = 0.765\frac{M}{2} \quad (10)$$

Then the optimal value of k is :

$$k = \frac{\sqrt{E_{fs}}}{\sqrt{E_{mp}}} \frac{\sqrt{N}}{\sqrt{2\pi}} \frac{M}{d_{toBS}^2} \quad (11)$$

Using Eqs. (10) and (11), we can obtain the energy E_{Round} dissipated during a round and thus we can compute the network lifetime R by Eq. (8).

5. WDDC PROTOCOL:

According to the Radio Energy Dissipation Model, the minimum required amplifier energy is proportional to the square of the distance from the transmitter to the destined receiver (Tx−Amplifier $\propto$ d²) [20]. So the transmission energy consumption will augment greatly as the transmission distance rises. It means that the CHs far from the BS must use much more energy to transmit the data to the BS than those close to the BS. Therefore, after the network operates for some rounds there will be significant difference between the energy consumption of the nodes near the BS and that of the nodes far from the BS.

In our approach, nodes with more energy than the other nodes and the nodes with less distance from the BS have more chance to be selected as a cluster-head for current round.

For this reason, we introduce new weighted probabilities for every type of nodes according to their residual energy and the average energy of the network in current round. We also take into consideration distances between nodes and the base station in order to favor nodes with more energy and nearest to the BS to become cluster heads.

The new probabilities are as follows:

$$P_{nrm} = \frac{p_{opt}E_i(r)}{(1+am)\bar{E}(r)} * (1-w)$$

$$P_{adv} = \frac{(1+a)\, p_{opt}E_i(r)}{(1+am)\bar{E}(r)} * (1-w)$$

$$(12)$$

Where w is the weighted factor that contains the notion of distances.

We suppose that after spreading the nodes in network field, the base station broadcasts a "hello" message to all the nodes at a given power level. Each node can compute its estimated distance (D_i) from the BS based on the received signal strength.

The average distance is given by:

$$D_{avg} = \frac{1}{N}\sum_{i=1}^{N} D_i \quad (13)$$

The value of D_{avg} can be approximated as:

$$D_{avg} \simeq d_{toCH} + d_{toBS} \quad (14)$$

Where:
- d_{toCH} The average distance between the node and the associate cluster head (figure 3).
- d_{toBS} The average distance between the cluster head and the base station (figure 3).

We assume that the nodes are uniformly distributed and the BS is located in the center of the field. Thus we can use equations (10) to calculate, d_{toCH}, d_{toBS} and finally D_{avg}.

We introduce the notion of far nodes and near nodes. However, nodes situated within the circle of radius equal to D_{avg} are considered near nodes and nodes situated in outside of the circle are considered far nodes (figure 4).

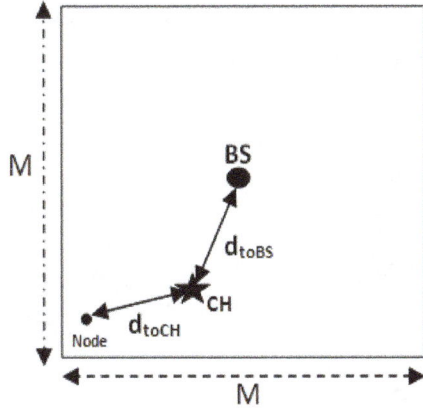

Figure 3. d_{toCH} and d_{toBS}.

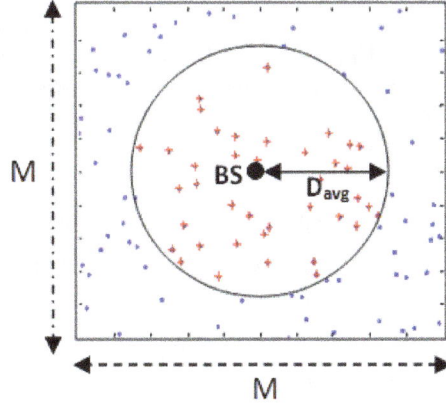

Figure 4. Near nodes (+) and far nodes (.)

The surface of the field of the network is obtained by:

$$A_1 = M * M \quad (15)$$

The surface of the circle is given by:

$$A_2 = \Pi * D_{avg}^2 \quad (16)$$

Where:

$$D_{avg} \simeq d_{toCH} + d_{toBS} \quad (17)$$

Using equations (10) we obtain:

$$D_{avg} \simeq \frac{M}{\sqrt{2k\pi}} + 0.765 \frac{M}{2} \quad (18)$$

The comparison gives:

$$A_2 < A_1$$

Finally, the weighted factor w is computed as follows:

$$w = A_2/A_1 \quad (19)$$

Furthermore, WDDC is dynamic because it divides network lifetime in two zones, strong zone and normal zone with size $m \times R$ and $R \times (1 - m)$ respectively (figure 5), in which it changes its behavior. So, in the first zone only advanced nodes become cluster heads. In the second zone the choice will become normal; it takes into consideration advanced and normal nodes. In both zones we use the new weighted probabilities mentioned in (12) and the probability threshold mentioned in (5).

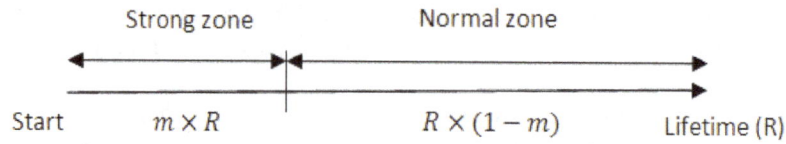

Figure 5. Strong zone and Normal zone

We notice that the size of zones is dynamic. It depends on the value of the fraction of the advanced nodes (m). The strong zone increases when m rises and vice versa for the normal zone; it decreases when m increases (figure 6).

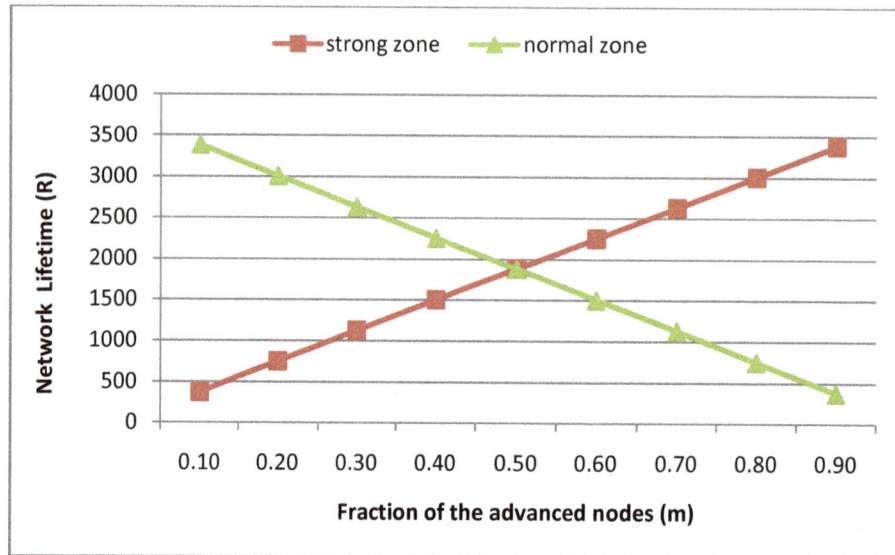

Figure 6. Sizes of strong zone and normal zone.

Communications between Cluster heads and member nodes

Like LEACH [3], after the cluster-heads are selected, the cluster-heads inform all sensor nodes in the network that they are the new cluster heads. And then, other nodes organize themselves into local clusters by choosing the most appropriate cluster-head (normally the closest cluster-head) (figure.7). Thereafter, the CH receives sensed data from cluster members according to TDMA schedule that was created and transmitted to them.

Communications between cluster heads and the base station.

Each node sends its data during their assigned transmission time to the respective associate cluster head. The CH node must keep its receiver on in order to receive all the data from the nodes in the cluster. When all the data is received, the cluster head node executes signal processing functions to compress the data into a single signal. When this phase is completed, each cluster head can send the combined data to the base station.

The consumed energy of cluster head CH_i is composed of three parts: data receiving, data aggregation and data transmission. Then:

$$E(CH_i) = m_i l E_{elec} + (m_i + 1) l E_{DA} + l(E_{elec} + \epsilon_{fs} d^2) \ (20)$$

Where: m_i is the sum of member nodes in associate cluster and $d=D_i$ the distance between CH and the BS (in this case we consider $< d_0$) (figure 7).

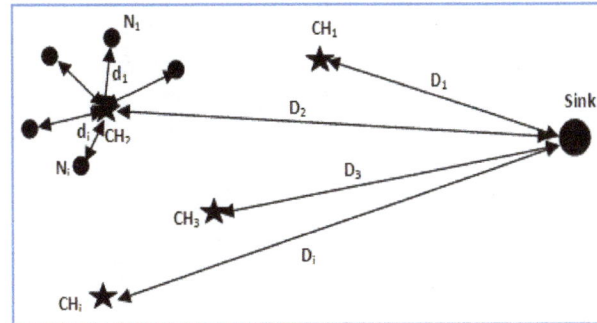

Figure 7. Distances between CH and the base station

6. SIMULATION RESULTS

We evaluate the performance of WDDC protocol using MATLAB. We consider a wireless sensor network with N = 100 nodes randomly distributed in a 100m × 100m field. We assume the base station is in the center of the sensing region. We ignore the effect caused by signal collision and interference in the wireless channel and we have fixed the value of d_0 at 70 meters like on DEEC.

The radio parameters used in our simulations are shown in Table 1. The protocols compared with WDDC include SEP and DEEC.

Table 1. Radio characteristics used in our simulations

Parameter	Value
E_{elec}	5 nJ/bit
ϵ_{fs}	10 pJ/bit/m^2
ϵ_{mp}	0.0013 pJ/bit/m^4
E_0	0.5 J
E_{DA}	5 nJ/bit/message
d_0	70 m
Message size	4000 bits
p_{opt}	0.1

Due to the heterogeneity factors, R is taken as 2×R (Since $\bar{E}(r)$ will be too large at the end from Eq.(7), thus some nodes will not die finally).
We define stable time as time until the first node dies (FND), and unstable time the time from the fist node dies until the last node dies. In other words, lifetime is the addition of stable time and unstable time.

We define also HNA (half of nodes alive) as the half of the total number of nodes that have not yet expended all their energy. All protocols remain useful during this period but after 50% of nodes die, the network becomes completely unstable and protocols will become useless.

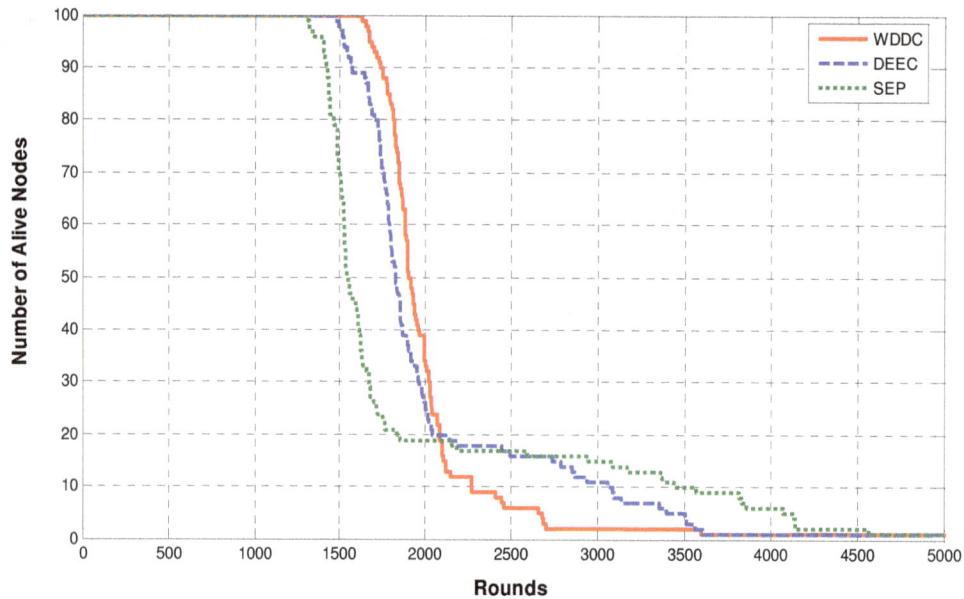

Figure 8. Number of nodes alive over time (m=0.2 and a=3)

According to the figure 8, we notice that the stable time of WDDC is large compared to that of SEP and DEEC. A longer stable time metric is important because it gives the end user reliable information of the sensing area, which extend the network lifetime. This reliability is vital for sensitive applications like tracking fire in forests.

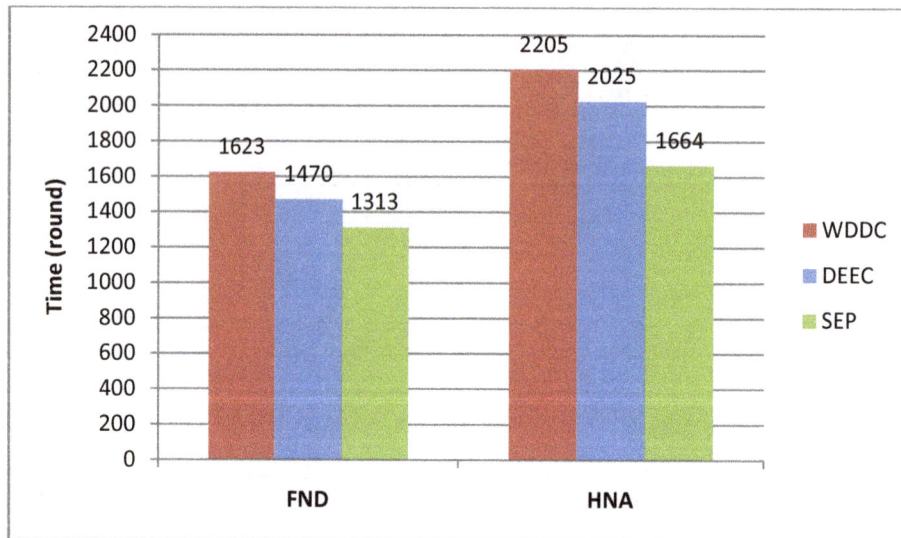

Figure 9. FND and HNA (m=0.2 and a=3)

Figure 9 shows the comparison between all nodes in terms of FND and HNA when m=0.2 and a=3. Obviously, we can remark that our protocol WDDC contains a larger period of stability time than SEP and DEEC, which increases the efficiency of the network.

We notice the same results for HNA. Therefore, WDDC performs better than the other simulated protocols. When half the number of nodes have expended all their energy, the network becomes inefficient.

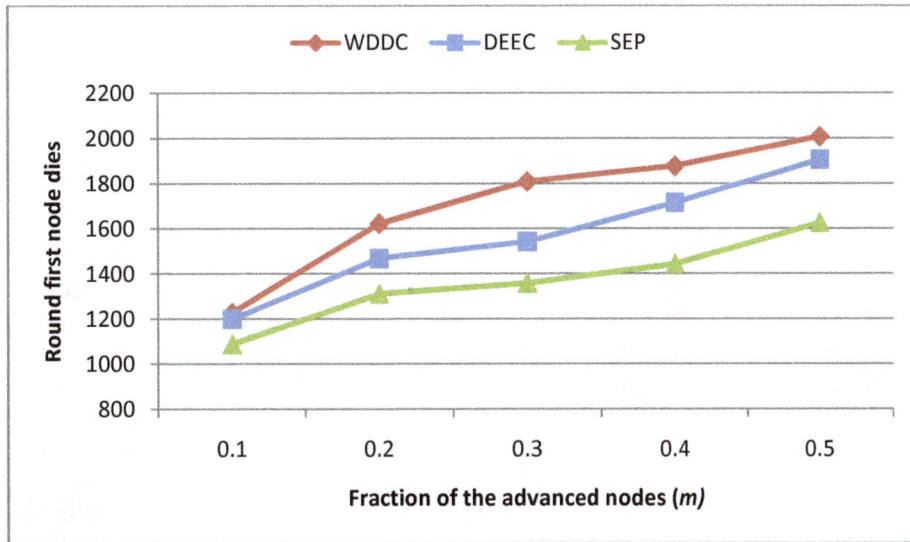

Figure 10. Round first node dies when m is varying.

Second, we run simulation for our proposed protocol WDDC to compute the round of the death of the first node when varying m, and we compare the results to SEP and DEEC protocols. We increase the fraction m of the advanced nodes from 0.1 to 0.5; Figure 10 shows the number of round when the first node dies. We observe that WDDC performs better than SEP and DEEC.

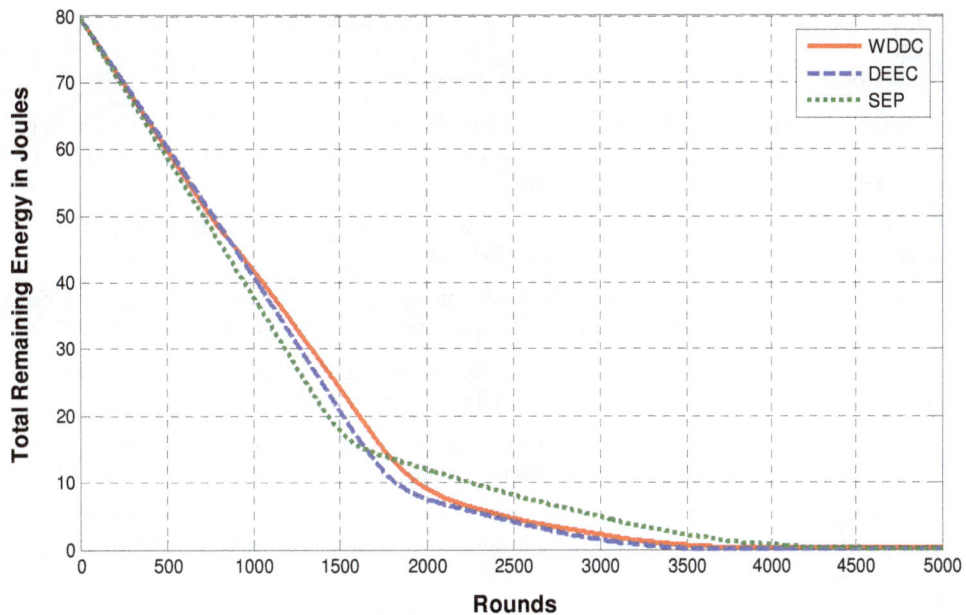

Figure 11. Total remaining energy over time of WDDC, DECC and SEP (m=0.2 and a=3)

Figure 11 shows the remaining energy over time for all simulated protocols and it reveals that WDDC consumes less energy in comparison to the others, which helps to extend the network lifetime. Here, approximately 7% of energy is saved at round 1500. This is because in our approach we took into consideration distances between nodes and the base station. Therefore, cluster heads situated far from the base station consume more energy than cluster heads situated near the base station, which saves the total energy of the network.

7. CONCLUSION

In this paper we have explained WDDC, a weighted dynamic distributed clustering protocol suitable for heterogeneous wireless sensor networks, and compared it to the SEP and DEEC protocols. WDDC is an energy-aware clustering protocol in which every sensor node separately elects itself as a cluster-head on the basis of probability of ratio of residual energy and average energy of the network and takes into consideration distances between nodes and the base station. Thus, nodes with less energy than the other nodes and the nodes with more distance from the BS have the smallest chance to be selected as a cluster-head for current round.

WDDC is dynamic, autonomous and more energy-efficient; it divides network lifetime in two zones, strong zone and normal zone, in which it changes its behavior so as to be more efficient. WDDC uses the two hierarchical levels concept which offers a better use and optimization of the energy dissipated in the network. Results from our simulations show that WDDC provides better performance for energy efficiency and network lifetime.

REFERENCES

[1] F. Akyildiz, W. Su, Y. Sankarasubramaniam, and E. Cayirici, "A survey on sensor networks", IEEE communications magazine 40 (8) (2002)102–114.

[2] J. N. Al-Karaki and A. E. Kamal, "Routing Techniques in Wireless Sensor Networks: A Survey", IEEE Journal of Wireless Communications, vol. 11, no. 6, Dec. 2004, pp. 6–28.

[3] W.B. Heinzelman, A.P. Chandrakasan, and H. Balakrishnan, "Energy efficient communication protocol for wireless microsensor networks", in Proceedings of the 33rd Hawaii International Conference on System Sciences (HICSS-33), January 2000.

[4] O. Younis, M. Krunz, and S. Ramasubramanian, "Node Clustering in Wireless Sensor Networks: Recent Developments and Deployment Challenges," IEEE Network (special issue on wireless sensor networking), vol. 20, issue 3, pp. 20-25, May 2006.

[5] A. Abbasi and M. Younis. "A survey on clustering algorithms for wireless sensor networks". Computer Communication, 30(14-15):2826–2841, 2007.

[6] S. Lindsey and C.S. Raghavenda, "PEGASIS: power efficient gathering in sensor information systems", in Proceeding of the IEEE Aerospace Conference, Big Sky, Montana, March 2002.

[7] O. Younis, and S. Fahmy, "HEED: A hybrid, energy-efficient, distributed clustering approach for ad hoc sensor networks", IEEE Transactions on Mobile Computing 3 (4) (2004) 660–669.

[8] D. Kumar, T. C. Aseri and R.B. Patel, "EEHC: Energy efficient heterogeneous clustered scheme for wireless sensor networks", Elsevier, Computer Communications 32 (2009) 662–667.

[9] G. Smaragdakis, I. Matta and A. Bestavros, "SEP: A Stable Election Protocol for clustered heterogeneous wireless sensor networks", in Second International Workshop on Sensor and Actor Network Protocols and Applications (SANPA 2004), 2004.

[10] V. Mhatre, and C. Rosenberg, "Design guidelines for wireless sensor networks: communication, clustering and aggregation", Ad Hoc Network Journal 2 (1) (2004) 45–63.

[11] M. Ye, C. Li, G. Chen and J. Wu, "EECS: an energy efficient cluster scheme in wireless sensor networks", in IEEE International Workshop on Strategies for Energy Efficiency in Ad Hoc and Sensor Networks (IEEE IWSEEASN2005), Phoenix, Arizona, April 7–9, 2005.

[12] A. Depedri, A. Zanella and R. Verdone, "An energy efficient protocol for wireless sensor networks", in: Autonomous Intelligent Networks and Systems (AINS 2003), Menlo Park, CA, June 30–July 1, 2003.

[13] L. Qing, Q. Zhu and M. Wang, "Design of a distributed energy-efficient clustering algorithm for heterogeneous wireless sensor networks". ELSEVIER, Computer Communications 29, pp 2230-2237, 2006.

[14] W. B. Heinzelman, A. P. Chandrakasan and H. BalaKrishnan, "An Application-specific Protocol Architecture for Wireless Microsensor Networks," IEEE Transactions on Wireless Communications, 1, No. 4, pp. 660–670, 2002.

[15] V. Loscri, G. Morabito and S. Marano, "A Two-Levels Hierarchy for Low- Energy Adaptive Clustering Hierarchy (TL-LEACH)". In 0-7803-9152-7/05/20.00 2005 IEEE.

[16] K. Latif, M. Jaffar, N. Javaid, M. N. Saqib, U. Qasim and Z. A. Khan. "Performance Analysis of Hierarchical Routing Protocols in Wireless Sensor Networks" NGWMN with 7th IEEE International Conference on Broadband and Wireless Computing, Communication and Applications (BWCCA 2012), Victoria, Canada, 2012.

[17] J. Y. Chang and P. H. Ju "An efficient cluster-based power saving scheme for wireless sensor networks". EURASIP Journal on Wireless Communications and Networking 2012 2012:172.

[18] R. Meelu and R. Anand "Performance Evaluation Of Cluster-Based Routing Protocols Used In Heterogeneous Wireless Sensor Networks" International Journal of Information Technology and Knowledge Management. Volume 4, No. 1, pp. 227-231, January-June 2011.

[19] S. Bandyopadhyay and E.J. Coyle, "An Energy Efficient Hierarchical Clustering Algorithm for Wireless Sensor Networks" in: Proceeding of INFOCOM 2003, April 2003.

[20] A. Khadivi and M. Shiva, "FTPASC: A Fault Tolerant Power Aware Protocol with Static Clustering for Wireless Sensor Networks", Proc. of IEEE Int. Conf. on Wireless and Mobile Computing, Networking and Communications, Montreal, Canada, Jun. 2006, pp. 397-401.

10

Cross-Layer Design Approach with Power Consciousness for Mobile Ad-Hoc Networks

Jhunu Debbarma[1], Sudipta Roy[2], Rajat K. Pal[3]

[1]Department of Information Technology, Triguna Sen School of Technology,
Assam University, Silchar , Assam 788011
[1]jhunudb@gmail.com
[2]Department of Information Technology, Triguna Sen School of Technology,
Assam University, Silchar , Assam 788011
[2]sudipta.it@gmail.com
[3]Department of Information Technology, Triguna Sen School of Technology,
Assam University, Silchar , Assam 788011
[3]pal.rajatk@gmail.com

ABSTRACT

The protocols used in mobile ad-hoc networks are based on the layered architecture. The layered approach is highly rigid and strict since each layer of the architecture is only concerned about the layers immediately above it or below it. Recent wireless protocols rely on significant interactions among various layers of the network stack. A cross-layer design (CLD) introduces stack wide layer interdependencies to optimize network performance. The CLD use the state information flowing throughout the network stack to adapt their behavior accordingly. In this paper, CLD based architecture is proposed, where the objective is to provide a solution for power conservation, congestion control, and link failure management. The link quality is determined by the received signal strength at the physical layer. The channel interference, contention and RTS/CTS packets of the MAC layer are used to determine the transmitting power and ensure the Quality of Service at the application layer.

KEYWORDS

Cross-layered design architecture, Optimization parameters, Power conservation, Signal strength.

1. INTRODUCTION

A mobile network is a group of mobile nodes that are equipped with wireless receiver and transmitter using antennas. As the nodes are vastly mobile, the network topology is unpredictable over time and varies actively. An ad-hoc network is very much deployable in this situation and without the need of any central administration. A mobile network is a group of wireless nodes that spontaneously build up independent networks without any fixed infrastructure or centralized administration [1]. For the purpose of communication among the nodes, the nodes need to perform packet routing. All the nodes in the mobile ad-hoc network (MANET) cooperatively maintain the network connectivity. The applications of MANET have wide range of network requirements along with different energy constraints for different network nodes. These requirements must be fulfilled despite of varying link characteristics on every hop, traffic, varying topology, and high mobility. One of the most critical issues of ad-hoc wireless network is that the activities of the nodes are power constrained since the nodes are powered by batteries. The present mobile ad-hoc wireless network protocol is based on layered approach, i.e., TCP/IP model. Each layer in this model is operated and designed independently, with interfaces among the layers. The interfaces are independent of the individual network constraints and applications. This paradigm of the interfaces has greatly simplified the network design and has contributed to robust, scalable protocols of the internet.

The objectives of routing algorithms in ad-hoc networks are based on optimization of multiple parameters instead of concentrating only on minimization of number of hops. Energy efficiency is one of the parameters to be optimized as the nodes have limited energy. In order to achieve that goal, vertical communication amongst the different layers of the protocol stack is required and this can be incorporated by cross-layer architecture. In this approach, different layers share useful information related to routing strategy to reduce the communication overhead and thus minimizing energy consumption of the participating nodes [2]. Some functions of the ad-hoc wireless network like mobility management, energy management, Quality of service (QoS), security, and cooperation cannot be implemented in a single layer of the network protocol. It is possible to implement these functions by exploiting and combining mechanisms of all the layers of the network protocol. A possible way to implement these functions is to avoid the rigid layering in which the protocols in each layer are developed in isolation but rather within an integrated and hierarchical framework that takes advantage of the interdependencies among them. The current ongoing debate among ad-hoc network researchers is cross-layered versus legacy-layered architectures.

In order to achieve desired optimization goal, there is need for information flow among different layers of the protocol stack which is termed as cross-layer design (CLD) approach. It relies on the interactions among layers of the network stack; see Figure 1. Cross layering can provide significant performance benefits though it is proved that the layered design has been one of the key elements of the success of Internet. The layers can share locally available information and this will improve the performance.

Application Layer	Energy management	Quality of service	Security and cooperation	Mobility management	Group Communication, Service Locations
Transport Layer					Transport Layer Protocols
Network Layer					TCP/IP routing, Addressing, Forwarding
MAC Layer					Framing, Error Detection and Control, Congestion
Link Layer					Antennas, MAC, Bluetooth, Power Control, 802.11, Hyper LAN.

Figure 1. MANET functions sharing between different layers through Cross Layer Design.

The different characteristics of the existing CLD architecture are enlisted as given below:

a) CLD involves the combinations of layers physical-MAC-network, MAC-network, network-transport only.

b) It provides individual solution for power conservation, energy minimization, flow control, congestion control, and fault tolerance.

c) Only the local link information from its MAC layer is used by the congestion avoidance algorithm.

d) There is high and expensive overhead.

The above mentioned features have certain drawbacks. There is still no work done on complete integration of MAC-network-transport layers. The local information from the MAC layer is not sufficient to replicate the network situation when the whole network becomes unstable. There is still no complete solution for power conservation, energy minimization, flow control, congestion control, and fault tolerance. Only individual solutions are there for these problems.

Due to high mobility of the nodes, there is always a high chance for frequent change of topology. To accommodate the dynamic topology and to facilitate communication in multi-hop fashion, reactive protocols are available. The Ad-hoc On-demand Distance Vector (AODV) is a reactive protocol that creates route to the destination only when the sender node has data to transmit by initiating a route discovery mechanism and maintains it until it is required by the source [3]. The source node initiates route establishment by broadcasting Route Request (RREQ) packet to its neighbours and waits for the Route Reply (RREP) packet from the destination or intermediate nodes that have fresh route information to the destination. A new CLD is proposed in this paper to provide a solution for unidirectional link failure management, reliable route discovery, and power conservation.

In view of all these, the paper is organized as follows. Section 2 presents the related works done in cross-layering design. Section 3 discusses the proposed cross-layer design architecture. Section 4 describes the simulation results, and the paper concluded in Section 5.

2. RELATED WORKS

A. J. Goldsmith *et al.* have identified that cross-layer approach to network design can increase the design complexity [4]. The layered protocol is useful in allowing designers to optimize single layer design without complexity and concerning other layers. The cross-layer design must consider the advantages of the layering keeping some form of separation among the layers. Each layer is identified by certain parameters that are to be shared by the layers just above or below it. The parameter sharing of the layers assists in determining the operation modes that are suitable for application conditions, network, and current channel situation.

S. Shakkottai *et al.* have discussed that Layer Triggers (predefined signals) are the basic cross-layer design implementation that provide quantifiable performance improvements by attaining compatibility through the extension of layered approach [5]. The example of Layer Trigger is Transmission Control Protocol (TCP) with Explicit Congestion Notification (ECN). The ECN mechanisms have an advantage to TCP by showing the differences between congestion loss and wireless channel related loss. TCP with ECN also avoids delays and packet loss, thereby improving the performance of the network.

L. Chen *et al.* have discussed the design of cross-layer congestion control and scheduling for wireless ad-hoc networks [6]. The scheduling constraint is formulated earlier by considering multi-commodity flow variables and resource allocation in networks with fixed wireless channels. The resource allocation problem resulted to three sub-problems: routing, scheduling, and congestion control.

B. Ramachandran *et al.* have discussed about a simple CLD between physical layer and MAC layer for power conservation based on transmission power control [7]. The carrier sense multiple access with collision avoidance of IEEE 802.11 is integrated with the power control algorithm. The exchange of Request-To-Send (RTS) / Clear-To-Send (CTS) control signal is used to piggyback the information to enable the sender node to discover the minimum power requirement to transmit the data.

An Adaptive Link-Weight (ALW) routing protocol is proposed by A. N. Al-Khwildi *et al.* [8]. This protocol selects an optimum route based on low delay, long route time, and available bandwidth. Cross-layering technique is used in which the ALW routing protocol is integrated with the application and physical layer. The proposed design allows applications to convey preferences to the ALW protocol to override the default path selection mechanism.

Premalatha *et al.* have discussed about the design challenges for energy constraint ad-hoc wireless network [9]. The full CLD architecture tries to exploit protocol design and layer interdependencies to optimize the overall network performance. In this case, control information is continuously flowing top-down and bottom-up in the protocol stack. An adaptive routing may be developed based on traffic, network, and current link condition. The application layer can utilize a notion of soft QoS by adapting the underlying network condition.

S. Mahlnecht *et al.* have proposed the use of explicit signaling to minimize the impact of mobility and link disconnection [10]. The explicit signaling includes route failure notification and route reestablishment notification from the intermediate nodes to notify the sender TCP about the disruption and to establish a new route.

X. Xia *et al.* have discussed that layer triggers are not sufficient to fix ad-hoc networks performance problem due to TCP-IP-MAC interactions [11]. Two-link-level mechanisms, link-RED, and adaptive spacing is introduced to improve TCP efficiency; hence a joint design of the TCP protocols and MAC protocols are essential.

M. Conti *et al.* have discussed that the protocols belonging to different layers can cooperate by sharing the network status information but at the same time maintaining the separation of layers for protocol design [12]. The proposed solution has the advantage of balanced cross-layer design. The cross-layering is limited to parameters and implemented through data sharing called network status, which is a shared memory that every layer can access. Interlayer cooperation is obtained by variable sharing and the protocols are still implemented in each layer.

3. PROPOSED CROSS-LAYER DESIGN

An approach is made to design a cross-layer architecture that is aimed at providing a combined solution for link failure management and power conservation.

a. To address the link failure problem, the received signal strength from the physical layer can help to determine the link quality. The links with low signal strength are discarded from the route selection [13].

b. To address congestion control, the channel interference and contention of the nodes can be estimated and notified to the application layer. This estimation of the MAC layer can be utilized by the application layer and the transmission rate can be adjusted accordingly, to avoid congestion.

c. To address the power conservation, the MAC layer RTS/CTS packet exchange can be used. The minimum required power can be estimated and accordingly the application layer can adjust the transmitting power.

3.1 Link Failure Management

The signal strength of the received signal can be estimated at the physical layer. This information is transferred to the MAC layer along with the signal strength information. The MAC layer uses this information for making calculations, later it is passed to the routing layer along with routing control packet. In the routing layer, the information is stored in the neighbour table (or routing table) and it is used in some decision making process. The IEEE 802.11 is reliable MAC protocol and it assumes fixed maximum transmission, since RTS must reach every exposed node and every CTS must reach every hidden node to avoid collision.

3.1.1 Power Consciousness for Energy Conservation

The nodes are having limited power and storage capacity, so power conscious cross-layer design is essential to save battery. A sender node while sending the RTS packet also attaches its transmission power. The receiver node measures the signal strength while receiving the RTS packet using the following relationship as shown in Eqn. (1).

$$T_R = T_S(\alpha/4\pi d)^2 S_T S_R \tag{1}$$

Here α is the wavelength of the carrier signal, d is the distance between the sender node and the receiver node. S_T is the unity gain of sending nodes omni-directional antennas and S_R is the unity gain of receiving nodes omni-directional antennas. T_S is the sender nodes transmission power and T_R is the received signal power at the receiving node.

The receiving node calculates the path loss experienced as shown in Eqn. (2).

$$Path\_loss = T_R - T_S \tag{2}$$

The minimum required transmission power P_{min} of the node is calculated by Eqn. (3).

$$P_{min} = L \times (Path\_loss + X_{th}) \tag{3}$$

Here L is the multiplying factor that provides marginal hike in minimum required transmission power to withstand against the effect of interferences on packet reception. X_{th} is the receiver threshold, the minimum received power essential for proper signal detection.

There are a set of protocols available for power control in mobile ad-hoc networks based on the common power approach [14]. These protocols are complex and have been analyzed that the variable range transmission power is a better approach than the common power.

In this paper, power control is also introduced to the RTS/CTS packets based on the received signal strength. When a source node wants to transmit data, it initiates the AODV routing protocol by broadcasting the RREQ packet to the neighbour nodes and the RREP packet is received from the intermediate nodes via the shortest route and then enters it in their routing table about the next hop to which the later data packets are needed to be forwarded.

For power conservation, the RREP packet is identified by an identifier (id) at the MAC layer and its signal strength information is obtained from the physical layer. The nodes that receive the AODV's RREP packet, compute two parameters— (i) path loss experienced using Eqn. (2) and (ii) minimum required transmission power using Eqn. (2) and Eqn. (3). The P_{min} and the next destination node information are stored in the routing table.

The proposed CLD works as follows:

a)   The nodes that send the RTS would refer to the routing table for the details of the minimum required transmission power.

b)   The sender node would then tune its transmission power and also inserts this value as an extra field in the RTS packet.

c)   The receiver node, on receipt of RTS, would tune its transmission power and replies back with CTS packet.

d)   Then the sender node would send the data with the requisite transmission power.

e)   The receiver node would also send the ACK with requisite transmission power.

This CLD involves the interaction of physical-MAC-routing layers. At the routing layer, the RREQ and RREP packets of the AODV routing protocol are transmitted with maximum transmission power so that bi-directionality of links, connectivity, and number of hops are unchanged. At the MAC layer, all the transmission sequences: RTS-CTS-DATA-ACK uses the minimum required power transmission level. The sender node on receipt of the ACK, calculates the path loss incurred using the currently used minimum transmit power value in its routing table to tackle high mobility. This adaptive transmit power updating mitigates unnecessary link/route failure due to the combined effect of power control and node mobility as transmission power is updated on per packet basis.

3.2 Unidirectional Link Rejection

The nodes in the ad-hoc networks are characterized by asymmetry links that means low-power nodes are able to receive from high power nodes but not vice versa. The AODV routing protocols has been designed for networks, with bidirectional links. The presence of asymmetric links become undetected and the RREP packet transmission along the reverse path fails.

In Figure 2, the route discovery process from node A to node B fails, since the RREP packet could not reach node A. This is due to the fact that, the AODV at the destination entertains the first received RREQ packet and does not reply to the RREQ packet via node C.

The AODV protocol allows only two RREQ retries. It fails to discover a route if there are more than two low-power nodes along the shortest route between the sender and receiver nodes.

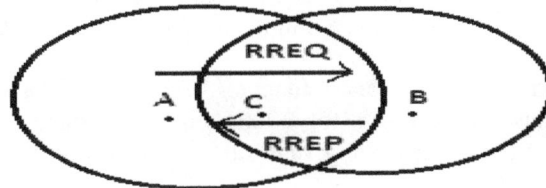

Figure 2. Route Reply failure due to low power nodes B, and C

To tackle the asymmetric links, different mechanisms are used as:

a) Periodic "Hello message" transmission when there is unidirectional link.

b) Black listing of nodes is done by storing the node where unidirectional link occurs and also to store the next hop of the failed RREP.

c) Reverse path search: In this scheme every node maintains multiple reverse paths while broadcasting RREQ. When RREP fails at a node the corresponding reverse path is erased and the RREP is retried along an alternate reverse path [15].

In this proposal, the unidirectional links are identified and rejected in the RREQ broadcast stage itself. If any bidirectional link exists, it is identified at the first RREQ packet broadcast. Whenever a node broadcasts the RREQ packet, it also includes the transmission power and antenna threshold value in the RREQ packet. On receipt of the RREQ packet by the receiver node, the path loss experienced by the RREQ packet is computed. It can detect if the link is bidirectional by comparing the sender node's antenna threshold value and the path loss value by Eqn. (4).

$$T_s > (Path\_loss + X_{th}) \qquad (4)$$

If so, then the link is bidirectional and the RREP packet may reach to the sender node. In this manner, the RREQ packet is processed as per route discovery process of AODV and the RREQ packet is broadcasted after replacing the transmission power field by its own transmission power value. If the transmission power is less, the RREQ packet is discarded and the unidirectional link is rejected in the RREQ forwarding phase itself.

3.3   Route Discovery

For reliable route discovery, the proposed CLD considers the received signal strength of RREQ to decide whether to forward or discard. The route discovery done in this manner is aimed to save resources, reduce route failure, and minimize routing overheads. The signal strength is compared to the defined fixed threshold value and decision is taken as to forward or discard [16].

In the proposed technique, the high mobility of nodes is taken into consideration by incorporating a parameter that decides if two nodes are becoming closer and moving apart. The received signal strength of RREQ is stored in the routing table against the address of the neighbouring nodes from which RREQ is received. The current value of received signal strength and the previous value are compared; if the current value is greater, that means the nodes are becoming closer else they are moving away from each other.

4. SIMULATION RESULTS

Network Simulator, NS2 is used for the experiments [17]. The simulation area is a square and the nodes are placed uniformly. Each node chooses a random point and moves towards that point with random speed chosen between minimum and maximum values.

The nodes use distributed coordination function of IEEE 802.11 standard with RTS/CTS extension. Simulations are executed for 1200s for three rounds at varying values. The parameters along with the corresponding values that are considered to carry on the simulation are enlisted in Table 1.

Table 1. Simulation parameters.

Parameters	Values
Radio frequency	2.5 GHz
Bandwidth	2 Mbps
Packet size	512 bytes
Inter-packet interval	0.3 s
Number of nodes	30
Network protocol	IP
Transport protocol	TCP
MAC protocol	IEEE 802.11
Routing protocol	AODV
Antenna gain	0 dBm
Receiver threshold	−80 dBm
Receiver sensitivity	−90 dBm
Grid area	500 m × 500 m
Speed	0 and 20 m/s
Traffic	Constant Bit Ratio (CBR)

Performance Metrics:

1. Average end-to-end delay: The end-to-end delay is averaged over all surviving data packets from the source to the destination.

2. Throughput: It is the number of packets received successfully.

3. Drop: It indicates the number of packets dropped.

4. Average Energy: It indicates the average energy consumption of all nodes sending, receiving and forwarding operation.

5. Average packet delivery ratio: It indicates the ratio of packets received successfully and the number of packets sent.

4.1 Energy Conservation

To analyze the properties for improving energy conservation with AODV routing protocol, the CLD was changed to transmit power control for all MAC packets. The CBR traffic was varied to change the offered load with randomly selected sender and receiver node. The amount of energy conservation in cross-layer design protocol (CLDP) ranges in between 10% to 25% as shown in Figure. 3 The modified cross-layer design protocol shows more collision than unmodified protocol (UMP) of IEEE 802.11 and AODV due to uneven power usages by the nodes; the low-powered nodes suffer from high interference caused by high-power nodes. The minimum power requirement can be estimated by the RTS/CTS packets of the MAC layer. The node sending RTS packet needs to refer the routing table and accordingly tunes its transmitting power as per the signal strength value. This value is also added as an extra field in the RTS packet such that the receiver can tune to this power while sending the CTS packet. In this way the collisions can be minimized.

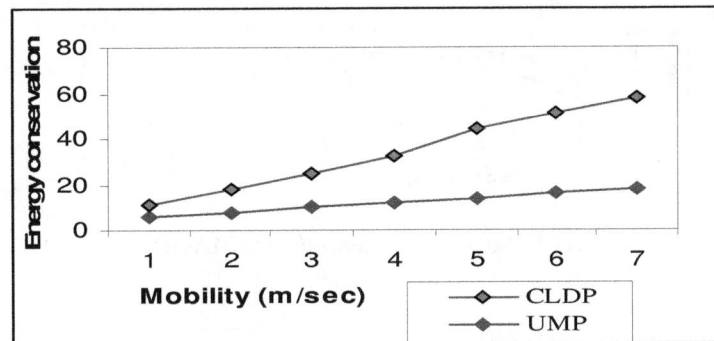

Figure 3. Energy conservation versus number of nodes for the cases of UMP and CLDP

4.2 Asymmetric Link Rejection

In the simulation model, all the nodes with 7 dBm are designated as high-powered with transmission range of about 250 m and nodes with 1 dBm are considered as low-powered with transmission range of 125 m. The simulation setup uses 25 nodes where 50% nodes are low-powered and the mobility of the nodes varies between 0 to 20 m/s. Four nodes are randomly selected as sender and receiver nodes, and the experiment is carried out for three times. In all the cases, it has been observed that there is improvement on packet delivery ratio of about 25-35% around the heterogeneously powered ad-hoc networks. There is reduced delay in route discovery. In heterogeneous environment, both the AODV and CLDAODV cause MAC collision. Link asymmetry causes low powered nodes to be hidden from the high powered nodes and this increases the number of collisions in the low powered communications. The simulation is considered with 50% low powered and 50% high powered nodes. In both situations the AODV and CLDAODV perform in the same manner due to dynamic network properties.

This implies that the CLDAODV's implementation does not degrade the performance in any form. The MAC collision is reduced to about 70-80% and the routing overhead is reduced to 75-85%. Hence, the proposed cross-layer design offers better performance since the unidirectional links are quickly identified and rejected before RREQ is broadcasted by the sender node.

4.3 Route Discovery Simulation Results

The AODV protocol with fixed threshold value is independent of the node's speed; so it is not justified for all speed values. In the proposed cross-layer protocol, AODV protocol is modified to tackle the situation of node mobility by considering the threshold values. The signal variant is fixed to −75 dBm in AODV and the modified AODV (MAODV) uses the set of values {−81, −80.5, −80, −78, −75 dBm} and it actually depends on the speed of the nodes in the range of 0-25 m/s.

The graphs in Figures 4(a)-4(d) depict the effect of mobility. There is improvement of Packet delivery ratio, average end-to-end delay and number of transmission in case of cross-layer designed AODV (CLDAODV) to the normal AODV protocol. The number of collisions is more in case of AODV than CLDAODV.

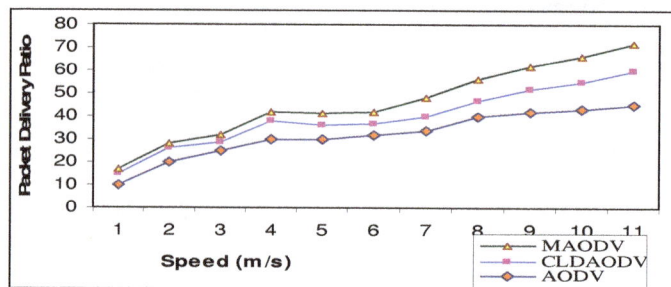

Figure 4(a): Packet delivery Ratio Versus speed(m/s) for MAODV, CLDAODV and AODV.

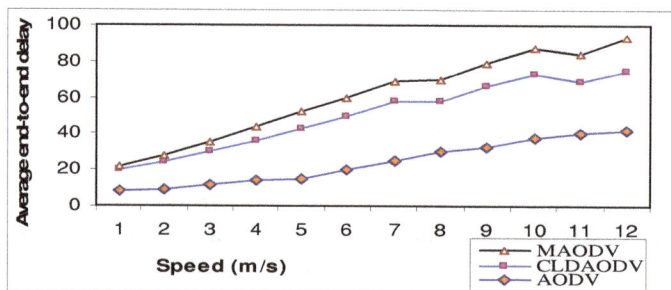

Figure 4(b). Average end-to-end delay versus speed(m/s) for MAODV, CLAODV and AODV.

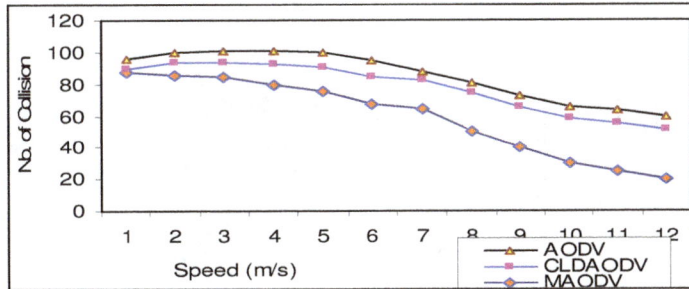

Figure 4(c). No. of collisions versus speed for MAODV, CLAODV and AODV

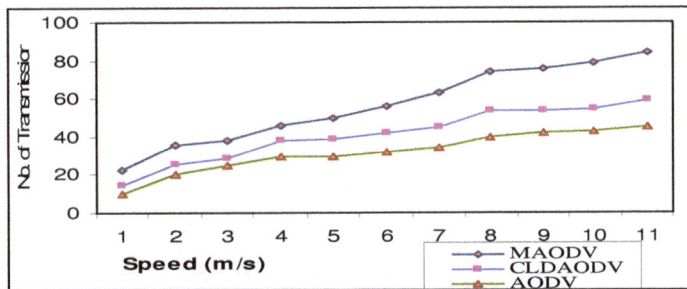

Figure 4(d). No of Transmission versus speed for MAODV, CLAODV and AODV

The MAC layer RTS/CTS packet exchange, help to estimate the minimum required power. The signal strength is obtained form the physical layer and this information is used by the routing layer.

Figures 5(a)-5(d) depict the effects of node density. Improvement is seen in the packet delivery ratio, average end-to-end delay and number of transmission. The collision rate is reduced in the CLDAODV.

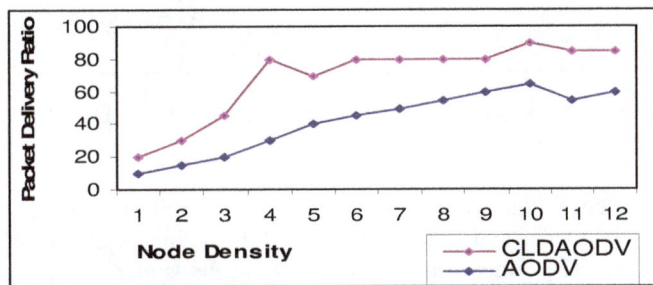

Figure 5(a). Packet delivery Ratio versus Node density for CLDAODV and AODV.

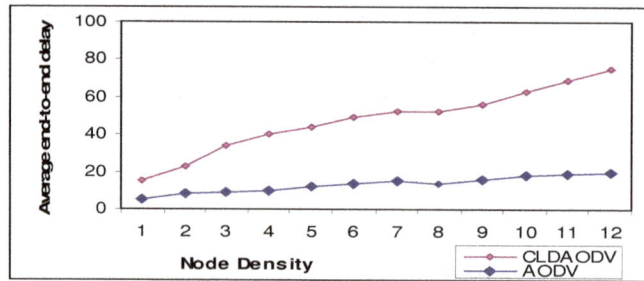

Figure 5(b). Average end-to-end delay versus Node Density for CLDAODV and AODV.

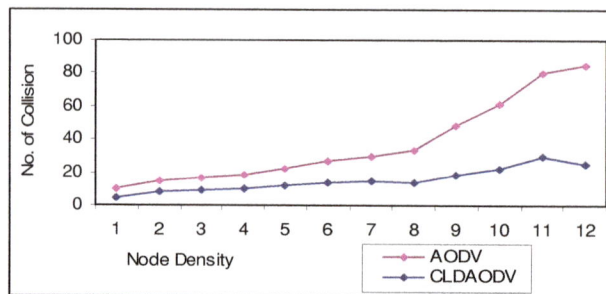

Figure 5(c). No. of collision versus Node Density for   CLDAODV and AODV.

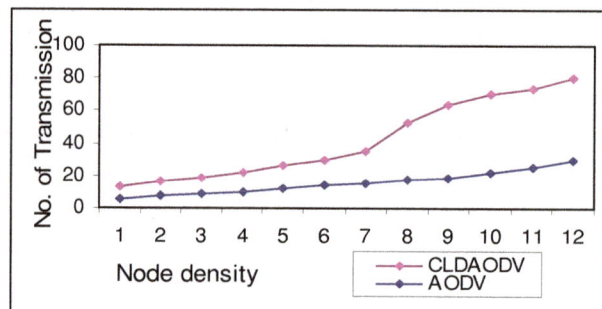

Figure 5(d).No. of transmission versus Node Density for CLDAODV And AODV.

The results as depicted in Figures 6(a)-6(d) show that the modified AODV protocol adaptively considers the threshold value and result into reduced delay, increased packet delivery ratio, and reduced route failure. The imposed threshold value on the signal strength affects the network connectivity. There is also improvement in routing overhead reduction due to reduced route failures. The cross-layer design need to be invoked in high density networks for better performance. The Modified AODV (MAODV) minimizes the number of hops when compared to the AODV with fixed variant. Hence the perfornance is improved in terms of increased packet delivery and reduced delay.

6. CONCLUSION

The high mobility and heterogeneous nature of the ad-hoc network results in collisions. The proposed cross-layer design is aimed to provide a solution for unidirectional link failure management, reliable route discovery, and power conservation. The link quality can be predicted by the received signal strength from the physical layer. The links having low signal strength can be discarded from the route selection. From the MAC layer, the minimum power required can be estimated by performing RTS/CTS packet exchange. Based on this, the application layer can readjust the transmission rate, to avoid collision.

One of the effective methods to reduce collision is to accompany the cross-layer design to achieve greater network capacity and spatial reuse. The proposed cross-layer design makes the AODV routing protocol to survive with heterogeneously powered ad-hoc networks by identifying and rejecting the asymmetric links at the RREQ broadcast stage itself. The most important fact is the network designers who must list down the conditions under which cross-layer design would improve the performance. To make accurate assessment of the state of the network efficient mechanisms need to be built into the protocol stack.

REFERENCES

[1]   IEEE standard 802.11, Part 11: Wireless LAN Medium Access Control and Physical Layer Specifications, 1999.

[2]   V. Srivastava and M. Motani, "Cross-Layer Design: A Survey and the Road Ahead", IEEE Communications Magazine, Vol. 43, No. 12, pp. 112-119, 2005.

[3]   C. E. Perkins, E. M. Royer, and S. R. Das, "Ad-hoc On-demand Distance Vector (AODV) Routing", IETF MANET Working Group, IETF RFC 3561, Jul. 2003.

[4]   A. J. Goldsmith and S. B. Wicker, "Design Challenges for Energy-Constraint Ad-Hoc Wireless Networks", IEEE Wireless Comm., Vol. 9, No. 4, pp. 8-27, 2002.

[5]   S. Shakkotai, T. S. Rappaport, and P. C. Karlsson, "Cross-Layer Design for Wireless Networks", IEEE Communications Magazine, Vol. 41, pp. 74-80, Oct. 2003.

[6]   L. Chen, S. H. Low, M. Chiang, and J. C. Doyle, "Cross-layer Congestion Control, Routing and Scheduling Design in Ad-Hoc Wireless Networks", Proc. of 25th IEEE International Conference on Computer Communications, pp 1-13, Apr. 2006.

[7]   B. Ramachandran and S. Shanmugavel,   "Received Signal Strength based Cross-Layer Designs in Mobile Ad-Hoc Networks", IETE Technical Review, Vol. 25. No. 4, pp. 192-200, 2009.

[8]   A. N. Alkhwildi, S. Khan, K. K. Loo, H. S. Al-Raweshidy, "Adaptive Link with Routing Protocols using Cross-Layer Communication for MANET", ISSN: 1109-2742, Issue 11, Vol. 6, Nov. 2007.

[9]   J. Premalatha, P. Balasubramanie, and C. Venkatesh, "Cross-Layer Design to Improve QoS in Mobile Ad-Hoc Networks", Journal of Mobile Communication, Vol. 2, No. 2, pp. 52-58, 2008.

[10]   S. Mahlknecht, S. A. Madani, and M. Roetzer, "Energy Aware Distance Vector Routing Scheme for Data Centric Low-Power Wireless Sensor Networks", IEEE Communications Magazine, Vol. 40, pp. 70-76, Oct. 2005.

[11]   X. Xia, Q. Ren, and Q. Liang, "Cross-Layer Design for Mobile Ad-Hoc Networks: Energy, Throughput, and Delay Aware Approach", Proc. of IEEE Conference on Wireless Communications and Networking, Vol. 2, pp. 770-775, 2006.

[12]   M. Conti, G. Maselli, and G. Turi, "Cross-Layering in Mobile Ad-Hoc Network Design", IEEE Computer Society, pp. 48-51, Feb. 2004.

[13]   V. Kawadia and P. R. Kumar, "A Cautionary Perspective on Cross-Layer Design", IEEE Wireless Communications, pp. 3-11, Feb. 2005.

[14]   V. Kawadia and P. R. Kumar, "Principles and Protocols for Power Control in Wireless Ad-Hoc Networks", IEEE Journal on Selected Areas in Communications, Part I, Vol. 23, No. 1, pp. 78-88, 2005.

[15]   M. K. Marina and S. R. Das, "Routing Performance in the Presence of Unidirectional Links in Multi-Hop Wireless Networks", Proc. of ACM Mohi-Hoc, pp. 12-23, 2002.

[16]   B. Ramachandran and S. Shanmugavel, "Reliable Route Discovery for Mobile Ad-Hoc Networks: A Cross-Layer Approach", Proc. of IETE International Conference on Next Generation Networks, pp. 26.1-26.6, Mumbai, 2006.

[17]   NS2 User Manual, http://www.isi.edu/nsnam/ns.

11

ENERGY CONSUMPTION ESTIMATION IN CLUSTERED WIRELESS SENSOR NETWORKS USING M/M/1 QUEUING MODEL

Reza Rasouli[1] and Mahmood Ahmadi[2] and Ali ahmadvand[3]

[1]Department of Information Technology,Science and Research branch,
Islamic Azad University Kermanshah, Iran
r.rasouli63@gmail.com
[2] Department of Computer Engineering, Faculty of Engineering,
University of Razi, Kermanshah, Iran
m.ahmadi@razi.ac.ir
[3] Department of Electronic and Computer,
Islamic Azad University, Qazvin Branch, Qazvin, Iran
ahmadvand@yahoo.com

ABSTRACT

Wireless Sensor Network (WSN), consists of a large number of sensor nodes. Each sensor node senses environmental phenomenon and sends the sensed data to a sink node. Since the sensor nodes are powered by limited power batteries. Energy efficiency is a major challenge in WSN applications. In this paper, we propose an analytical model for energy consumption estimation in clustered WSNs using M/M/1 queuing model. The model can be used to investigate the network performance in terms of average energy consumption. We also propose a power minimization scheme to reduce energy consumption of sensor nodes by reducing the number of transitions between the idle and active states of sensor nodes according to the number of data packets in the queue. Our analytical model, based on the main parameters such as the average energy consumption and the number of packets (jobs), significantly reduces energy consumption. Simulation, support the validity of the proposed approach.

KEYWORDS

Wireless Sensor Networks (WSNs), Clustered Network, Ambient Network, Energy Consumption Model, M/M/1 Queuing Model.

1. INTRODUCTION

WSNs, consist of a large number of small sensor nodes with sensing, computation, and wireless communications capabilities [1]. In these applications sensors are usually remotely deployed in large quantities and operated autonomously [2]. Limited lifetime of the sensors and possible damages lead to the need for a strategic management [1,2].

Amongst the wide range of applications, disaster management is one of the situations in which WSNs can be applied [3].

In these situations such as earthquakes and flood, WSNs can be used to selectively map the affected regions directing emergency response units to Similarly, in military situations, WSNs can be used in surveillance missions and detect moving targets, detection and monitoring applications [1].

One of the critical issues in such applications is represented by the limited availability of power within the network and hence consumption power is vital [4], and leads to an increase in network lifetime. Several techniques have been introduced for saving energy, such as the use of energy efficient routing and switching between sleep/active modes for sensors [5].

For the first category many routing, power management [6, 7, 8], and data dissemination protocols have been specifically designed for WSNs where energy awareness is an essential design issue. Routing protocols [9, 10] in WSNs might differ depending on the application and network architecture.

WSNs [11, 12, 13, 14] generates a large amount of data in which has to be aggregated at various levels.

In this paper, we consider a clustered WSNs model that sensor nodes are uniformly distributed in the field. A Cluster Heads (CHs) node at the centre collects data from the sensor nodes. A sink node at the centre collects data from the CH nodes.

We propose an analytical model, using M/M/1 queuing model, to estimate total energy consumption of sensor nodes, the parameter, used to control the overall energy and battery power are rationalized to provide best possible solution [15]. Our analytical model based on the main parameters such as the average energy consumption and the number of packets (jobs) is that significantly reduces energy consumption.

In our model, we assume that cluster members (CM) periodically sense the environment and send bach the sensed data to the CH node. The arrival of data packets to sensors is assumed to follow a Poisson process with mean arrival rate (λ_{CM}) per node and with policy arrival FCFS.

For more energy saving, we propose a power minimization scheme by which the power consumption in all sensor nodes in the network is optimized by reducing the number of transitions between the idle and active states of the sensor nodes . Focuses working on reducing the number of transitions between idle and active state is, that will lead to reduced energy consumption. Finally validate our analytical model using simulations.

Our contributions in this paper are as follows:

1 – Introducing of an analytical model for energy consumption estimation in clustered wireless sensor networks using M/M/1 queuing model.

2 - To compute energy consumed for transmitting data to the sink node, we model each node such as ordinary sensor node (CMs), CH nodes and sink node based on simple M/M/1 queue.

3 – Suggesting a power minimization scheme to reduce energy consumption of sensor nodes by reducing the number of transitions between the idle and active states based on the number of data packets in the queue.

4 - Validate our analytical model using simulations.

This paper is organized into sections. The sections provide the information about the parts and modules of the research undertaken by the current statements. Section 1 provides the introduction. Section 2 includes the background of the energy conservation techniques in WSNs, Section 3 deals with the proposed clustered network architecture. In Section 4, we present system model and the proposed analytical model to estimate the energy consumption in clustered based WSNs. The proposed work may be extended with the simulation results which validate the proposed model in section 5. The conclusion is stipulated in Section 6.

2. RELATED WORK

We review some related work on energy conservation techniques in wireless sensor networks. WSNs have many challenges in use and deployment including limited power resources and also limited memory and communication capabilities [16].

However, the most important challenges in the WSNs are energy consumption because battery capacities of sensor nodes are limited and replacing them in many applications are impractical. Sensor nodes consume most energy for data transmission and reception [17].

Sensor network lifetime due to limited power considerably depending on the battery power of the sensor nodes and if the energy consumption of nodes is reduced, thus increasing the network lifetime.

R.Maheswar and R.Jayaparvathy in [20] developed an analytical model of a clustered sensor network using M/G/1 queuing model. The authors have analyzed the system performance in terms of energy consumption and the mean delay.

One of the best techniques for minimizing energy consumption in wireless sensor networks is switching between the active and sleep state [19]. In this technique, when a sensor node has packet to send, it turns on and switches to active state and sends packet. When sensor has not any packet to send, it switches to sleep state and turns off. If the number of transitions between the active and sleep state is high, energy consumption is high. Also in idle state, sensor nodes consume less energy than active state.

R. Maheswar and R. Jayaparvathy [20,21] introduce an energy minimization technique using BUSY and IDLE states where the energy consumed is minimized based on queue threshold using M/M/1 queuing model.

By Fuu-Cheng Jiang and et al [22], an analytical model based on the M/G/1 queuing model to reduce energy consumption by reducing the average time to access media is competition.

3. CLUSTERED NETWORK ARCHITECTURE

A general backbone of network that we are going to analyze it energy consumption is depicted in Figure 1. Figure 1, demonstrates a heterogeneous clustered network. Similar to Yarvis work [18], a three stage architecture is used. In this architecture, three different types of nodes with different hardware capabilities and battery power are used. The sensor nodes with higher hardware capabilities and more battery power compared to other sensor nodes act as cluster heads (CHs) and sink nodes in the network. In this architecture, the top level contains only the sink node that receives sensed data and analyze them. The second level consists of cluster heads nodes (CHs). CHs task aggregate data and send it to a sink node. The lowest level contains sensor nodes (CMs) that can only connect to its one hop cluster head. In this network, we have three different types of nodes, sink node, cluster head(CHs), and sensor nodes (CMs), as shown in Figure 1. Instance application of the architecture are presented in this paper in the following two cases can be cited. In disaster management situations such as earthquakes, flood, Etc. WSNs can be used to selectively map the affected regions directing emergency response units to survivors. Also in military situations, WSNs can be used in surveillance missions and can be used to detect moving targets, detection and monitoring applications [1].

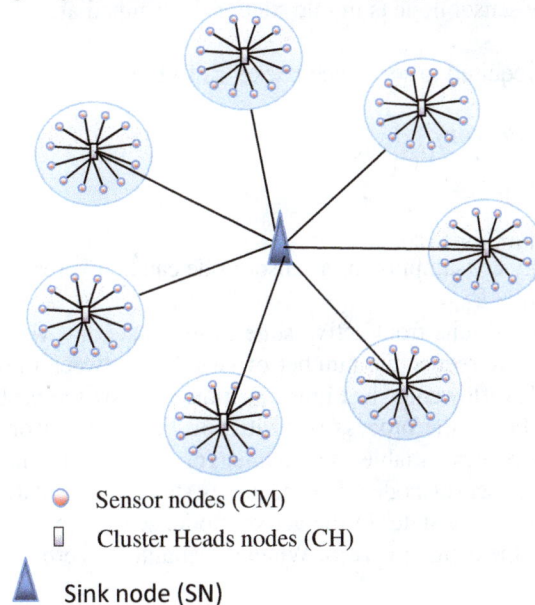

○ Sensor nodes (CM)

▫ Cluster Heads nodes (CH)

▲ Sink node (SN)

Figure 1- Clustered Network Architecture.

In our WSN model, the following assumptions are made.
- All sensor nodes (CMs) are identical

- All cluster head nodes (CHs) are identical
- No channel contention
- The arrival of data packets to sensors is assumed to follow a Poisson process with mean arrival rate (λ_{CM}) per node
- The buffer capacity of the cluster head node and sink node is infinite
- Each cluster has only one cluster head.
- The number of ordinary sensor nodes in each cluster are the same.

4. ENERGY CONSUMPTION ANALYTICAL MODEL

4.1. Energy consumption in sensor nodes

For our analytical model, the following notations are used.

That include: λ_{CM} , μ_{CM}, U_{CM} , Π_0 , N , C , M , E_{Tx} , E_{idle} that respectively is equal to   mean arrival rate per sensor nodes (CMs) , mean service rate in sensor nodes (CMs) ,utilization of the sensor nodes (CMs) , probability that the sensor node is in idle state , mean number of packets in sensor node per unit time , Number of clusters , number of sensor node per cluster , energy consumption for transmit one data packet , energy consumption in idle state in sensor node .

The steady state balance equations obtained for the analytical model according to the M/M/1 queuing model which are given by equations (1) to (6).

Utilization of the sensor nodes (CMs) is determined as:

$$U_{CM} = \lambda_{CM} \Big/ \mu_{CM} \tag{1}$$

Mean number of packets in the sensor node (N) is determined as:

$$N = U_{CM} \Big/ (1 - U_{CM}) \tag{2}$$

The probability that the sensor node is in idle state is determined as:

$$\pi_0 = 1 - U_{CM} \tag{3}$$

The amount of energy required to send each packet as follows:

$$E_{TX} = \text{Transmission Power} \Big/ \mu_{CM} \tag{4}$$

Where

$$\mu_{CM} = \text{Band Width} \Big/ \text{Packet Size} \tag{5}$$

Now, the average energy consumption of a sensor node can be expressed as:

$$PW_{CM} = N * E_{TX} + \pi_0 * E_{Idle} \tag{6}$$

In analytical model, transitions from active state to idle state and vice versa consume most of the energy in WSNs. If we reduce the number of switches between these states, we can reduce energy consumption. The flowchart in Figure 3, indicates switching between active and idle mode. The flowchart shows, the process of switching between sensor nodes in two state the active and idle mode. Counter variable as a counter for counted the number of jobs that can be entered into the queue of sensor nodes. Sensor nodes stays in idle state until receive B packets and then switches to the active state. In the active mode, sensor nodes transmitting or receiving or processing jobs until the counter is zero. When the counter is zero, the sensor nodes switch to idle state.

In this analytical technique sensor node stays in idle state until receive B packets and then switches to the active state, it can reduce the number of switches between the states.

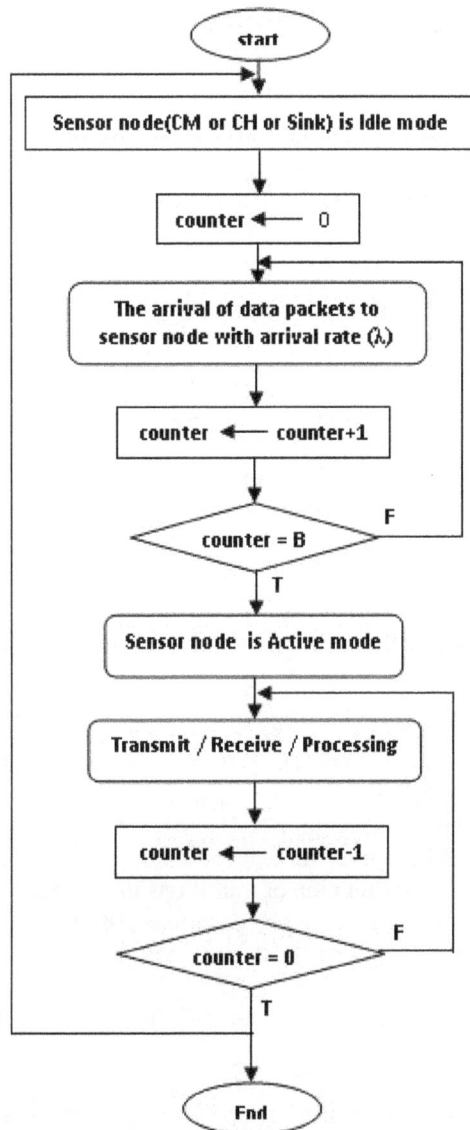

Figure 3- The flowchart switching between active and idle mode.

To calculate the energy consumption of transitions in a single sensor node, we define the following parameters: T, E_{st} , N_t , PW_T , N_t, B_{CM} , P_{LO} , T_{st} , that respectively is equal to Total time, Start up energy consumption in transceiver , Number of transitions during T seconds , Total energy consumption for transitions at during T seconds and   transmit per sensor node , Buffer size of sensor node ,The power consumption of the circuitry including the synthesizer and the VCO , Time required to start up all the transceiver components .

The steady state balance equations are given by equations (7) to (10):

The number of transitions during T seconds can be expressed as;

$$N_t = (T * \lambda_{CM})/B_{CM} \tag{7}$$

Time required to start up all the transceiver components can be expressed as;

$$E_{st} = P_{LO} * T_{st} \tag{8}$$

Now, the total energy consumption of a sensor node can be expressed as:

$$PW_{SW1} = N_t * E_{st} \tag{9}$$

Finally, the total energy consumption for M sensor nodes in the per cluster is:

$$E_{CMs} = M * (PW_{CM} + PW_{SW1})$$ (10)

4.2. Energy consumption in cluster head nodes

The steady state balance equations obtained for the analytical model according to the M/M/1 queuing model which are given by equations (11) to (17).

Mean arrival rate per CH nodes can be expressed as :

$$\lambda_{CH} = (M - 1) * \lambda_{CM}$$ (11)

Utilization of the CH node is determined as:

$$U_{CH} = \lambda_{CH} / \mu_{CH}$$ (12)

Mean number of packets in the CHs node N_1 is determined as:

$$N_1 = U_{CH} / (1 - U_{CH})$$ (13)

The probability that the CHs node is in idle state is determined as:

$$\pi_{0-CH} = 1 - U_{CH}$$ (14)

Energy required to receive a packet is:

$$E_{RX} = \text{Receive Power} / \mu_{CH}$$ (15)

The processing of data is only CHs nodes, thus energy required to processing data packet is:

$$E_p = \text{Processing Power} / \mu_{CH}$$ (16)

Now, the average energy consumption of a CHs node can be expressed as:

$$PW_{CH} = N_1 * (E_{TX} + E_{RX} + E_p) + (\pi_{0-CH} * E_{Idle-CH})$$ (17)

In this analytical technique CHs node stays in idle state until receive B_{CH} packets and then switches to the active state, it can reduce the number of switches between the states. To calculate the energy consumption of transitions in a single CH node:

The steady state balance equations are given by equations (18) to (19):

The number of transitions during T seconds can be expressed as;

$$N_{t1} = (T * \lambda_{CH}) / B_{CH}$$ (18)

Now, the energy consumption of a CH node for switching between the states can be expressed as:

$$PW_{SW2} = N_{t1} * E_{st}$$ (19)

Also, the total energy consumption for per CH nodes in the network is:

$$E_{CHs} = PW_{CH} + PW_{SW2}$$ (20)

Finally, the average energy consumption in per cluster in the network of equations (10) and (20) is obtained, which is equivalent to:

$$E_{Cluster} = E_{CHs} + E_{CMs}$$ (21)

Also, the total energy consumption for C cluster in the network is:

$$E_{Cluster-total} = C * E_{Cluster}$$ (22)

4.3. Energy consumption in sink node

The steady state balance equations obtained for the analytical model according to the M/M/1 queuing model which are given by equations (23) to (27).

Mean arrival rate per sink node can be expressed as:

$$\lambda_{Sink} = (C - 1) * \lambda_{CH}$$ (23)

Utilization of the sink node is determined as:

$$U_{Sink} = \lambda_{Sink} / \mu_{Sink}$$ (24)

Mean number of packets in the sink node N_2 is determined as:

$$N_2 = U_{Sink} \Big/ (1 - U_{Simk}) \tag{25}$$

The probability that the sink node is in idle state is determined as:
$$\pi_{0-Sink} = 1 - U_{Sink} \tag{26}$$
Now, the average energy consumption of a sink node can be expressed as:
$$PW_{Sink} = N_2 * (E_{TX} + E_{RX}) + (\pi_{0-Sink} * E_{Idle-Sink})) \tag{27}$$
Sink node only received incoming packets from the CH nodes and send to base station. Also sink node stays in idle state until receive B_{sink} packets and then switches to the active state, it can reduce the number of switches between the states. The steady state balance equations are given by equations (28) to (29):
The number of transitions during T seconds can be expressed as;
$$N_{t2} = (T * \lambda_{Sink})/B_{Sink} \tag{28}$$
Now, the total energy consumption of a sink node can be expressed as:
$$PW_{SW3} = N_{t2} * E_{st} \tag{29}$$
The total energy consumption for sink node in the network is:
$$E_{Sink} = PW_{Sink} + PW_{Sw3} \tag{30}$$
Finally, the total energy consumption in the network of equations (22) and (30), is obtained, which is equivalent to:
$$E_T = E_{Cluster-total} + E_{Sink} \tag{31}$$

For example, consider in a Clustered WSNs with 10 clusters and each cluster consists of 10 sensors, where mean arrival rate per sensor node of 0.1 ($\lambda_{CH} = 0.1$), mean service rate in sensor node of 0.5 ($\mu_{CH} = 0.5$), bandwidth of 10 kbps, packet size of 1600 bytes, E_{Idle} of 0.2 and transmission power of 0.4 joule, then we have following values:
The utilization of a sensor node is:
$$U_{CM} = \frac{\lambda_{CM}}{\mu_{CM}} = \frac{0.1}{0.5} = 0.2 * 100 = 20\%$$
The mean number of packets in sensor node is:
$$N = \frac{U_{CM}}{1 - U_{CM}} = \frac{0.2}{1 - 0.2} = 0.25$$
The probability that the sensor node is in idle state is:
$$\pi_0 = 1 - U_{CM} = 1 - 0.2 = 0.8$$
The energy required for sending a data packet is:
$$E_{TX} = \frac{\text{Transmission Power}}{\mu_{CM}} = \frac{\text{Packet size}}{\text{Band width}} * \text{Transmission Power} =$$
$$\frac{1600 * 8}{10 * 1000} 0.4 = 0.512$$
And the average energy consumption of a sensor node per unit time is:
$$PW_{CM} = N * E_{TX} + \pi_0 * E_{Idle} =$$
0.25 * 0.512 + 0.8 * 0.2=0.288
To calculate the energy consumption of transitions in a single sensor node, we have following values:
T=1000, $P_{LO} = 0.2$, $T_{st} = 0.1$, $B_{CM} = 10$
The number of transitions during 1000 seconds is:
$$N_t = \frac{(T * \lambda_{CM})}{B_{CM}} = \frac{1000 * 0.1}{10} = 10$$
Startup energy consumption in transceiver is:
$$E_{st} = P_{LO} * T_{st} = 0.2 * 0.1 = 0.02$$
Now, the total energy consumption of a sensor node can be expressed as:
$$PW_{SW1} = N_t * E_{st} = 10 * 0.02 = 0.2$$

Finally, the total energy consumption for M sensor nodes in the per cluster is:
$E_{CMs} = M * (PW_{CM} + PW_{SW1}) = 10 * (0.288 + 0.2) = 4.88$
Now, energy consumption in CH node:
Mean arrival rate per CH nodes can be expressed as :
$\lambda_{CH} = (M - 1) * \lambda_{CM} = (10 - 1) * 0.1 = 0.9$
The utilization of a CH node is:
$$U_{CH} = \frac{\lambda_{CH}}{\mu_{CH}} = \frac{0.9}{0.98} = 0.92 * 100 = 92\%$$
The mean number of packets in CH node is:
$$N_1 = \frac{U_{CH}}{1 - U_{CH}} = \frac{0.92}{1 - 0.92} = 11.5 \text{ packet}$$
The probability that the CH node is in idle state is:
$\pi_{0-CH} = 1 - U_{CH} = 1 - 0.92 = 0.08$
Energy required to receive a packet is:
Receive Power = Processing Power = 0.3,
$$E_{RX} = \frac{\text{Receive Power}}{\mu_{CM}} = \frac{0.3}{0.98} = 0.306$$
Energy required to processing data packet is:
$$E_P = \frac{\text{Processing Power}}{\mu_{CM}} = \frac{0.3}{0.98} = 0.306$$
Now, the average energy consumption of a CHs node can be expressed as:
$E_{Idle-CH} = 0.2$,
$PW_{CH} = N_1 * (E_{TX} + E_{RX} + E_P) +$

$$(\pi_{0-CH} * E_{Idle-CH}) = 11.5 * (0.512 + 0.306 + 0.306)$$

$$+(0.08 * 0.2) = 12.96$$
To calculate the energy consumption of transitions in a single CH node:
The number of transitions during T seconds can be expressed as:
$$N_{t1} = \frac{(T * \lambda_{CH})}{B_{CH}} = \frac{1000 * 0.9}{10} = 90$$
Now, the energy consumption of a CH node for switching between the states can be expressed as:
$PW_{SW2} = N_{t1} * E_{st} = 90 * 0.02 = 1.8$
Also, the total energy consumption for per CH nodes in the network is:
$E_{CHs} = PW_{CH} + PW_{SW2} = 12.942 + 1.8 = 14.742$
The average energy consumption in per cluster is:
$E_{Cluster} = E_{CHs} + E_{CMs} = 4.88 + 14.742 = 19.622$
Also, the total energy consumption for C cluster in the network is:
$E_{Cluster-total} = C * E_{Cluster} = 10 * 19.622 = 196.22$
Now, energy consumption in sink node :
Mean arrival rate per sink node can be expressed as :
$\lambda_{Sink} = (C - 1) * \lambda_{CH} = (10 - 1) * 0.9 = 8.1$
The utilization of sink node is:
$$U_{Sink} = \frac{\lambda_{Sink}}{\mu_{Sink}} = \frac{8.1}{8.88} = 0.91 * 100 = 91\%$$
The mean number of packets in sink node is:
$$N_2 = \frac{U_{Sink}}{1 - U_{Sink}} = \frac{0.91}{1 - 0.91} = 10.12 \text{ packet}$$
The probability that the sink node in idle state is:

$\pi_{0-\text{Sink}} = 1 - U_{\text{Sink}} = 1 - 0.91 = 0.09$

The average energy consumption of sink node can be expressed as:

$PW_{Sink} = N_2 * (E_{TX} + E_{RX}) + (\pi_{0-\text{Sink}} * E_{\text{Idle-Sink}}) =$

$$10.12 * (0.512 + 0.306) + (0.09 * 0.4) = 8.314$$

To calculate the energy consumption of transitions in sink node:

$$N_{t2} = \frac{(T * \lambda_{\text{Sink}})}{B_{\text{Sink}}} = \frac{1000 * 8.1}{20} = 405$$

Now, the energy consumption of sink node for switching between the states can be expressed as:

$PW_{SW3} = N_{t2} * E_{st} = 405 * 0.02 = 8.1$

The total energy consumption for per sink node in the network is:

$E_{Sink} = PW_{Sink} + PW_{SW3} = 8.314 + 8.1 = 16.414$

Finally, the total energy consumption in the network of equations is obtained,

$$E_T = E_{\text{Cluster-total}} + E_{\text{Sink}} = 196.2 + 16.414 = 212.634 \text{ joule}$$

5. SIMULATION MODEL

Now, we use simulations to validate our analytical model. We use simulator MATLAB for wireless sensor networks. The various network parameters and the power consumption parameters of Mica2 mote sensors used for the simulation model are shown in Table 1.

Table 1: Model and simulation parameter

Parameter	Value
Mean arrival rate per sensor nodes	0.0 1 to 0.1
Mean service time	0.5 sec
Number of sensor nodes per each cluster (M)	5 to 10
Number of Cluster (C)	10
Packet size	1600 b
Band width	10 Kbps
Transmission Power (Tp)	0.4 watt
Plo	0.01 watt
Tst	0.1 sec
Idle Power	0.1 watt
Receive Power (RP)	0.2 watt
Processing Power (PP)	0.2 watt
Time simulation	1000 sec

Simulations results are obtained for various scenarios by changing the mean arrival rate of sensor node and also changing the number of cluster heads. We measure the average energy consumption of all M sensor node and the average energy consumption of C cluster heads.

In Figure 4 –a, X-axis is equal to the mean arrival rate (Lambda) and Y-axis is equal to the average energy consumption in the ordinary sensor node. From Figure 4-a, it is results that the average energy consumption for per sensor node increases as mean arrival rate increases. Our analytical model is according simulations.

In Figure 4–b, X-axis is equal to the mean arrival rate (Lambda-CH) and Y-axis is equal to the average energy consumption in the CH nodes. From Figure 4-b, it is results that the energy consumption for per CH node increases as mean arrival rate increases.

In Figure 4–c, X-axis is equal to the mean arrival rate (Lambda-Sink) and Y-axis is equal to the average energy consumption in the sink node. From Figure 4-c, it is results that the energy consumption for sink node increases as mean arrival rate increases.

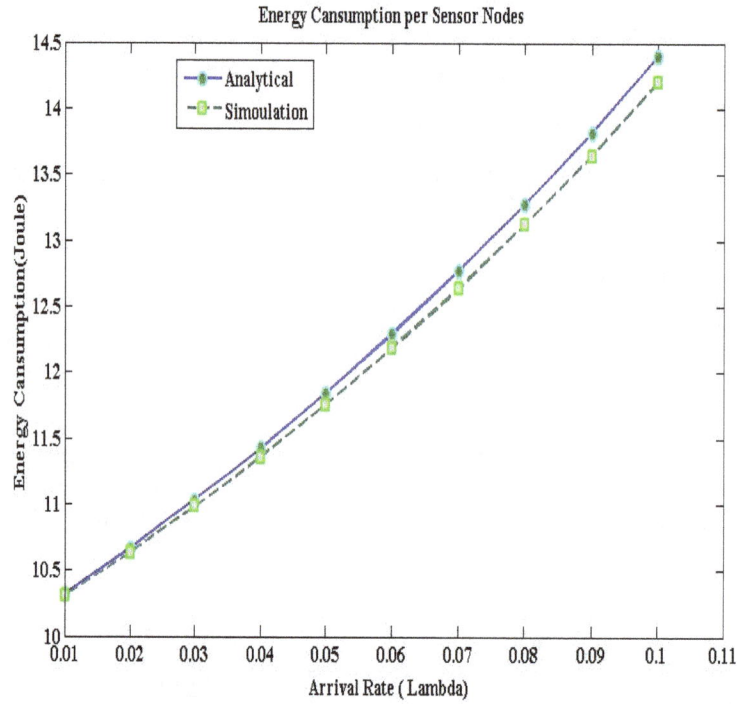

Figure 4-a- Mean arrival rate vs. energy consumption per sensor nodes (CM).

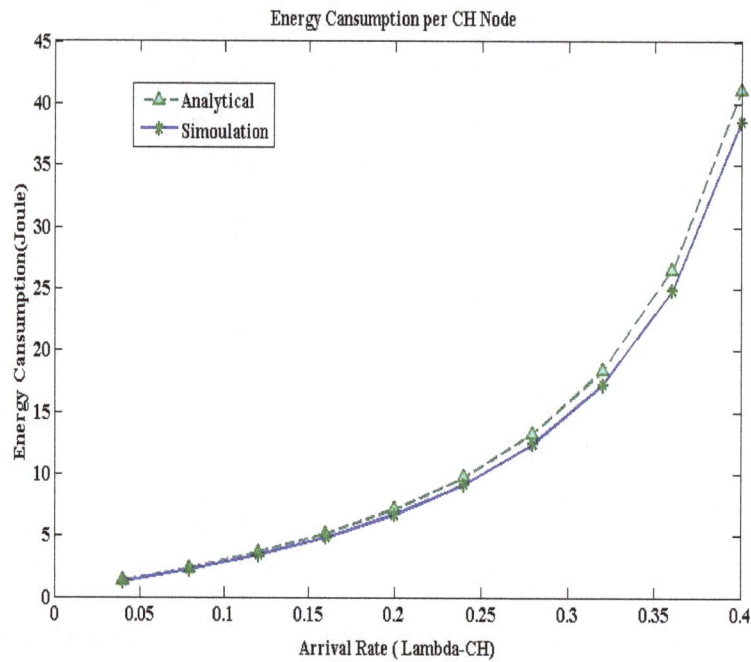

Figure 4-b- Mean arrival rate vs. energy consumption per CH node.

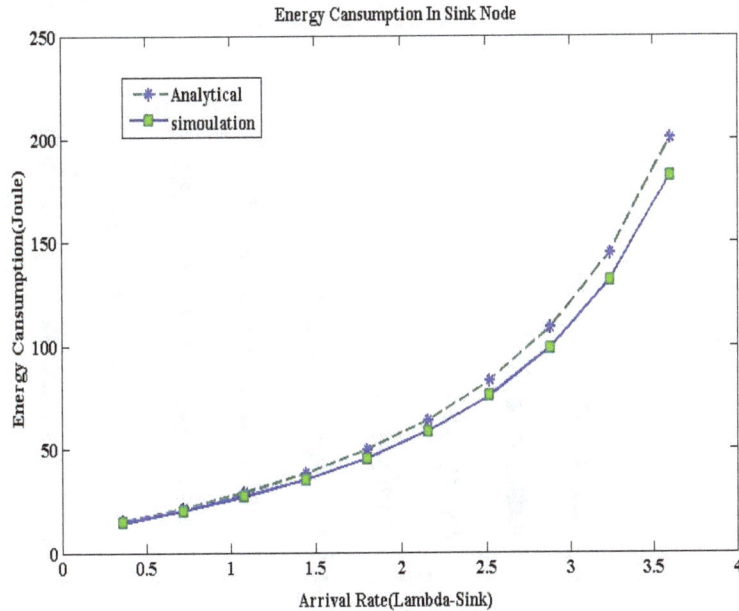

Figure 4-c- Mean arrival rate vs. energy consumption per   sink node.

In Figure 5 –a, X-axis is equal to the mean arrival rate (lambda) and Y-axis is equal to total energy consumption per all sensor nodes (CMs). From Figure 5-a, it is results that the average energy consumption for all sensor node increases as mean arrival rate increases because the number of packets that arrive to the sensor node increases. As you see of Figure 5-a, we calculated Energy consumption for all sensor nodes. The simulations were performed for 10 runs and a confidence interval of 99% was obtained.

In Figure 5 –b, X-axis is equal to Number of cluster and Y-axis is equal to total energy consumption per all CH nodes (CHs). Figure 5-b represents the energy consumption per all CH nodes with different number of clusters, for the average arrival rate is different. Our analytical model matches 99% with simulation result.

In Figure 5–c, X-axis is equal to the mean arrival rate (lambda) and Y-axis is equal to total energy consumption per sink node. Figure 5-c shows average energy consumption in sink node for various mean arrival rate of 0.01 to 0.1 packet per second. As shown in Figure 5-c, the average energy consumption in sink node increases as mean arrival rate increases.

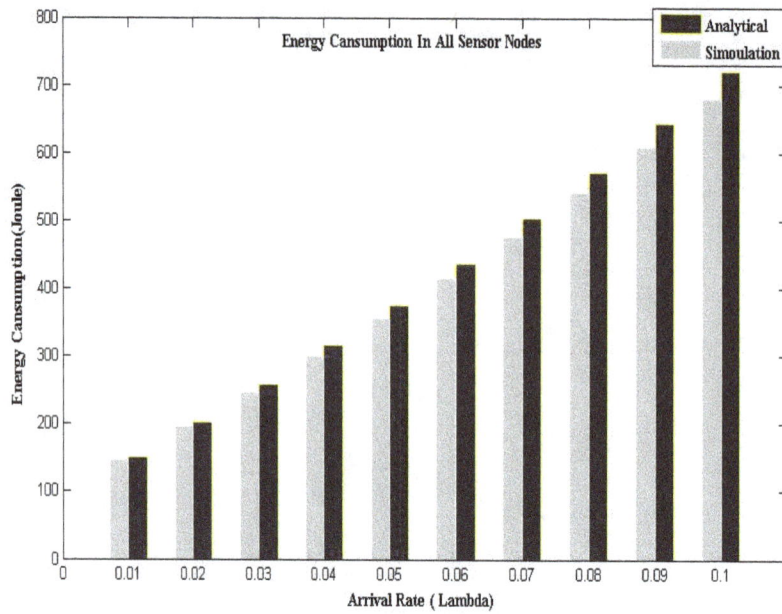

Figure 5-a - Mean arrival rate vs. total energ consumption per all sensor nodes (CMs).

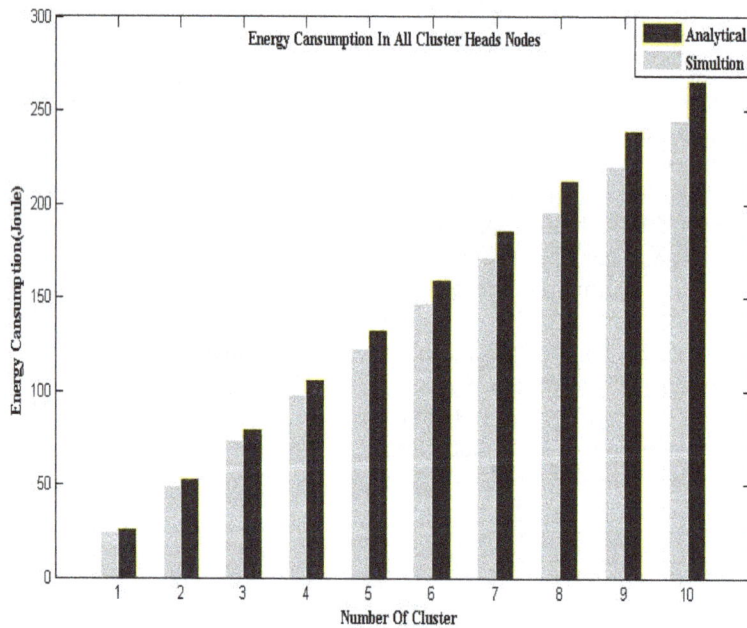

Figure 5-b . Number of cluster vs. total energy consumption per all CH nodes (CHs).

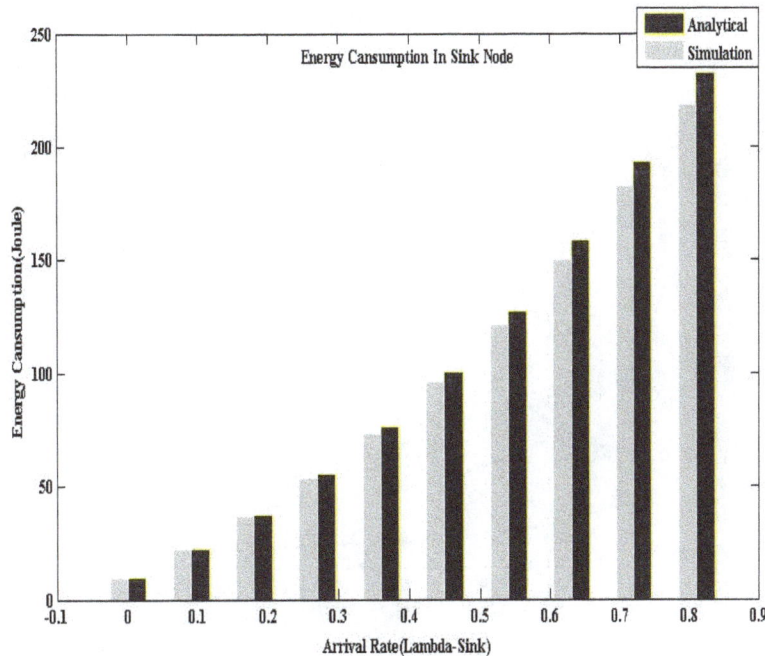

Figure 5-c - Mean arrival rate vs. total energy consumption in Sink node .

In Figure 6-a - X-axis is equal to B_{CM} value and Y-axis is equal to energy consumption for transitions during T=1000 seconds and with arrival rate of 0.1 to 0.4 packet per second in sensor node.Figure 6-a shows the result for T=1000 seconds. Form this Figure, we observe that energy consumption per sensor node with arrival rate of 0.1 to 0.4 packet per second decreases as the B_{CM} value increases. Figure 6-a, also shows that energy consumption increases as the mean arrival rate increases. From Figure 6-a we can conclude that, there are small differences between energy consumption for transitions for various mean arrival rate for $B_{CM} > 8$.

In Figure 6-b, X-axis is equal to B_{CH} value and Y-axis is equal to energy consumption for various number of sensor node per cluster in CH node.Figure 6-b shows that energy consumption per CH node with various number of sensor node per cluster decreases as the B_{ch} value increases. Figure 6-b, also shows that energy consumption increases as number of sensor node per cluster increases. From Figure 6-b we can conclude that, there are small differences between energy consumption for transitions for various number of sensor node per cluster for $B_{CH} > 8$.

In Figure 6-c, X-axis is equal to B_{Sink} value and Y-axis is equal to energy consumption for various number of cluster in Sink node. Figure 6-c shows that energy consumption in sink node with various number of CH node decreases as the B_{Sink} value increases. Figure 6-c, also shows that energy consumption increases as number of CH node increases. From Figure 6-c we can conclude that, there are small differences between energy consumption for transitions for various number of cluster for $B_{Sink} > 9$.

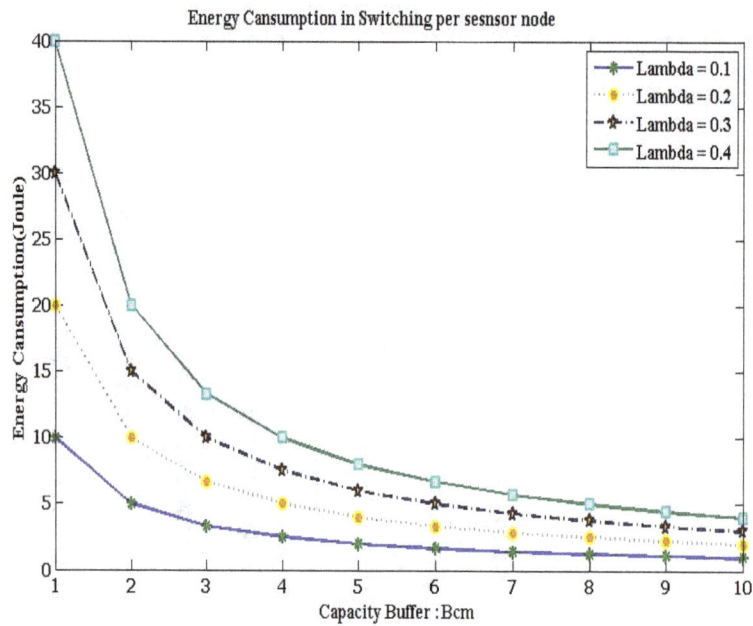

Figure 6-a -   value vs. energy consumption for   transitions during T=1000 seconds and with arrival rate of 0.1 to 0.4 packet per second in sensor node

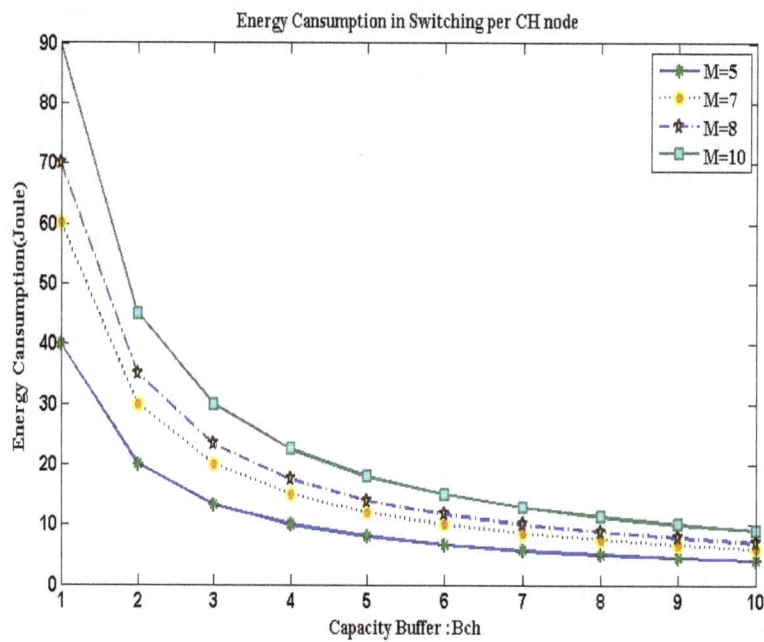

Figure 6-b-   value vs. energy consumption for   various number of sensor node per cluster in CH node

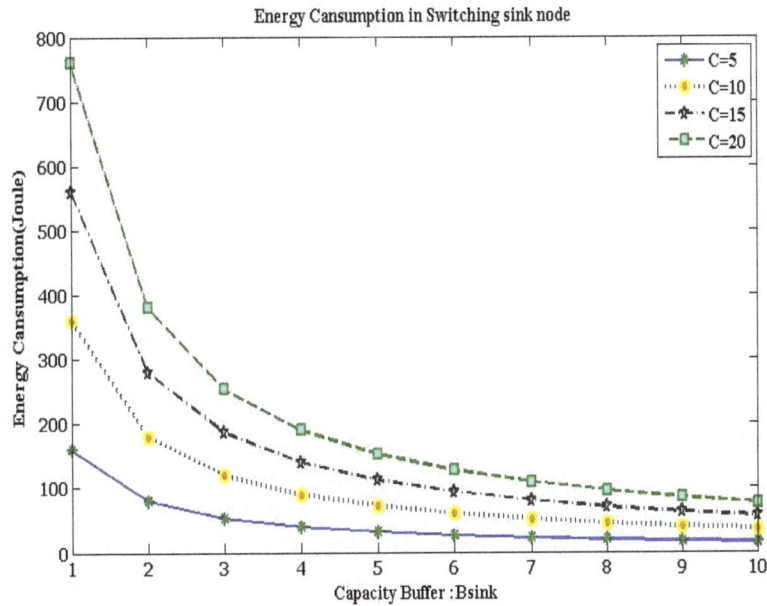

Figure 6-c -   value vs. energy consumption for   various number of cluster in Sink node

In Figure 7-a, X-axis is equal to Mean arrival rate and Y-axis is equal to total energy consumption (Et) in WSNs. Figure 7-a, shows total energy consumption for various mean arrival rate of 0.01 and 0.1 packet per second. As shown in Figure 7-a, the total energy consumption increases linearly as mean arrival rate increases.

In Figure 7-b, X-axis is equal to Number of cluster and Y-axis is equal to total energy consumption (Et) in WSNs. Figure 7-b, shows total energy consumption for various number of cluster of 1 to 10.

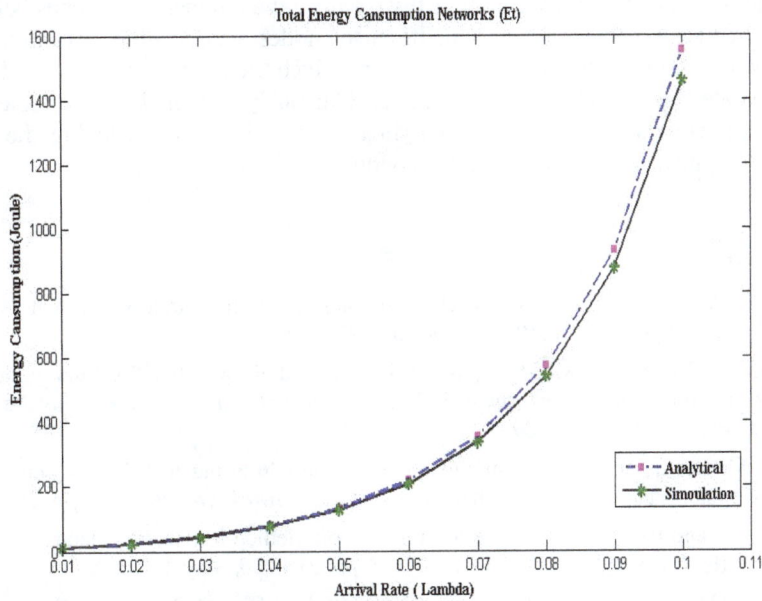

Figure 7-a - Mean arrival rate vs. total energy consumption (Et) in WSNs.

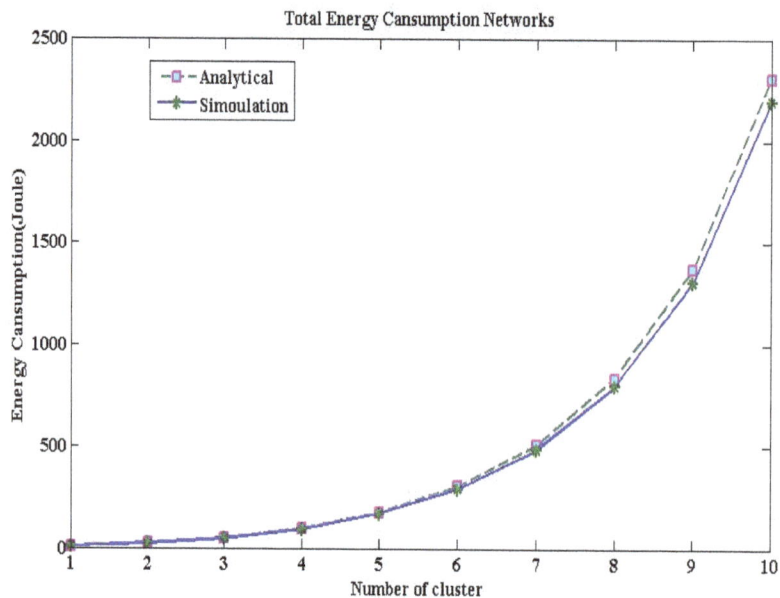

Figure 7-b - Number of cluster vs. total energy consumption (Et) in WSNs.

6. CONCLUSIONS

In this paper, we proposed a new analytical model to estimate the energy consumption in clustered WSNs using M/M/1 queuing model for all sensor nodes. Transitions from active state to idle state and vice versa consume most of the energy in WSNs. So we purpose have developed an analytical model for energy saving by reducing the number of transitions between idle sate and active state in all sensor nodes. If we reduce the number of switches between these states, we can reduce energy consumption. If sensor nodes stays in idle state until receive B packets and then switches to the active state, it can reduce the number of switches between the states. We validate this model using simulations. Our analytical model is suitable for delay-tolerant applications. The results of our analytical model show that reducing the number of transitions has a significant impact on energy saving.

REFERENCES

[1] A. A. Abbasi, M. Younis, "A survey on clustering algorithms for wireless sensor networks", *2007 Published by Elsevier B.V doi:10.1016/j.comcom.2007.05.024*

[2] O. Boyinbode , H. le , A. Mbogho , M. Takizawa and R. poliah , "A Survey on Clustering Algorithms for Wireless Sensor Networks", *13th International Conference on Network-Based Information Systems,PP 358-364, 2010*

[3] O. Younis, M. Krunz, and S. Ramasubramanian, "Node Clustering in Wireless Sensor Networks: Recent developments and deployment challenges," *IEEE Network, vol. 20, no. 3, pp. 20–25, 2006.*

[4] R. Jayaparvathy and R. Maheswar, "Power Optimization Method for Heterogeneous Sensor Network with Finite Buffer Capacity", *IJRTET, Vol. 3, No. 3, pp. 218-220, May 2010.*

[5] Z. Quan, A. Subramanian and A. H. Sayed, "REACA: An Efficient Protocol architecture for Large Scale Sensor Networks," IEEE Transactions on Wireless Communications, *vol. 6, no. 10, pp. 3846—3855 (2007).*

[6] E. Liu, Q. Zhang, and K. Leung, "Residual Energy-Aware Cooperative Transmission (React) in Wireless Networks ," *In Wireless and Optical Communications Conference (Wocc), pp. 1–6, May 2010.*

[7] K. Zarifi, A. Ghrayeb, and S. Affes, "Distributed   Beamforming for Wireless Sensor Networks with Improved Graph Connectivity And Energy Efficiency," *Signal Processing, IEEE Transactions, Vol. 58, No. 3, pp. 1904–1921, March 2010.*

[8] M. Mehrani, A. Shaeidi, "A Novel Energy Efficient, Distributed, Clustering Based Network Coverage Method for Enormous WSN", *Global Journal of Computer Science and Technology Volume 11 Issue 4 Version 1.0, pp. 59-65, March 2011.*

[9] S. K. Singh, M P Singh, D K Singh, "A Survey of Energy-Efficient Hierarchical Cluster-Based Routing in Wireless Sensor Network's", *570 Volume: 02, Issue: 02, pp: 570-580 (2010).*

[10] M. Z. Siam, M. Krunz, and O. Younis, "Energy-Efficient Clustering Routing for Cooperative Mimo Operation in Sensor Networks," *In Infocom, pp. 621–629, 2009.*

[11] G. Xin, "An Energy-Efficient Clustering Technique for Wireless Sensor Networks", *IEEE Conference, ISBN: 978-0-7695-3187-8, 25 July 2008*

[12] M. H. Anisi, A. H. Abdullah, S. A. Razak, "Energy-Efficient Data Collection in Wireless Sensor Networks", *Wireless Sensor Network, pp.329-333, October 2011.*

[13] X. Guan, L. Guan and X. Wang, "A Novel Energy Efficient Clustering Technique Based on Virtual Hexagon for Wireless Sensor Networks", *Volume 7, Issn 1349-4198, pp. 1891-1904, April, 2011.*

[14] S.K. Singh, M P Singh, and D K Singh, "Energy Efficient Homogenous Clustering Algorithm for Wireless Sensor Networks", *International Journal of Wireless & Mobile Networks (Ijwmn), Vol.2, No.3, August 2010.*

[15] S. Chaudhary, N. Singh, A. Pathak and A.K Vatsa , "Energy Efficient Techniques for Data aggregation and collection in WSN " , *International Journal of Computer Science, Engineering and Applications (IJCSEA) Vol.2, No.4, August 2012*

[16] I. Akyldiz, W.Su, Y. Sankarasubramanian and E. Cayirci, "A survey on sensor networks," *IEEE Commun. Mag., vol. 40, no. 8 , pp. 102-14, Aug. 2002*

[17] R.B. Patel, D. Kumar, and T.C Aseri, "EECDA: Energy-efficient Clustering and Data Aggregation Protocol For Heterogeneous Wireless Sensor Networks," *International Journal of Computers, Communication and Control, Romania, pp. 113-124 ,Vol. 6, 01, 2011.*

[18] M. Yarvis, N. Kushalnagar, H. Singh, A. Rangarajan, Y. Liu, and S. Singh,"Exploiting Heterogeneity in Sensor Networks", *Proceedings IEEE INFOCOM '05, vol. 2, Miami, F., pp. 878-890, Mar '2005.*

[19] J. Carle and D. Simplot-Ry, "Energy-efficient area monitoring for sensor networks". *IEEE Computer, vol. 37, no. 2, pp, Feb. 2004.*

[20] R. Maheswar and R. Jayaparvathy, "Performance Analysis of Clustered Sensor Networks using N-Policy M/G/1 Queueing Model", *European Journal of Scientific Research , Euro Journals Publishing, Inc. 2011*

[21] R. Jayaparvathy and R. Maheswar, "Power Control Algorithm for Wireless Sensor Networks using N-Policy M/M/1 Queueing Model", *IJCSE Vol. 02, No. 07, pp. 2378-2382, . August 2010.*

[22] F. C. Jiang , C. T. Yang , K. H. Wang, " Design Framework sleeping to Optimize Power Consumption Computing, and Latency Delay for Sensor Nodes using Min (N, T) Policy M/G/1 Queuing Models", *PP 344-351 , 2010 IEEE*

12

ENERGY HARVESTING WIRELESS SENSOR NETWORKS: DESIGN AND MODELING

Hussaini Habibu[1], Adamu Murtala Zungeru[2], Ajagun Abimbola Susan[3], Ijemaru Gerald[4]

[1,3]Department of Electrical and Electronics Engineering
Federal University of Technology Minna, Nigeria
[2,4]Department of Electrical and Electronics Engineering
Federal University Oye-Ekiti, Nigeria

ABSTRACT

Wireless sensor nodes are usually deployed in not easily accessible places to provide solution to a wide range of application such as environmental, medical and structural monitoring. They are spatially distributed and as a result are usually powered from batteries. Due to the limitation in providing power with batteries, which must be manually replaced when they are depleted, and location constraints in wireless sensor network causes a major setback on performance and lifetime of WSNs. This difficulty in battery replacement and cost led to a growing interest in energy harvesting. The current practice in energy harvesting for sensor networks is based on practical and simulation approach. The evaluation and validation of the WSN systems is mostly done using simulation and practical implementation. Simulation is widely used especially for its great advantage in evaluating network systems. Its disadvantages such as the long time taken to simulate and not being economical as it implements data without proper analysis of all that is involved ,wasting useful resources cannot be ignored. In most times, the energy scavenged is directly wired to the sensor nodes. We, therefore, argue that simulation – based and practical implementation of WSN energy harvesting system should be further strengthened through mathematical analysis and design procedures. In this work, we designed and modeled the energy harvesting system for wireless sensor nodes based on the input and output parameters of the energy sources and sensor nodes. We also introduced the use of supercapacitor as buffer and intermittent source for the sensor node. The model was further tested in a Matlab environment, and found to yield a very good approach for system design.

KEYWORDS

Wireless Sensor Networks (WSNs); Mathematical Analysis; Energy Harvesting; Simulation.

1. INTRODUCTION

Wireless sensor networks (WSNs) are large networks consisting of small sensor nodes (SNs), with limited computing resources used to gather , process data and communicate. A major challenge in a lot of sensor network applications requires long period of life for network survival, which leads to high consumption of energy. The small sensor nodes are devices driven by battery and due to its high energy demand, the conventional low-power design techniques and structure cannot provide an adequate solution [1]. Wireless sensor nodes normally run on disposable batteries, which have a finite operating life. Based on the application and availability of potential ambient energy sources, using energy harvesting techniques to power a wireless sensor node is a wonderful thing to do.

Wireless Sensor nodes have wide range of applications in our day to day activities. Ranging from a Bluetooth equipped chest band that convey human heart rate to a treadmill, wireless electrocardiograph (ECG) temporarily connected to communicate human cardiac activity to a doctor, Zigbee equipped smart meter that monitors energy usage in a household and provides feedback to the user for decision making [2]. In general, wireless sensor nodes applications include structural monitoring, industrial monitoring, security, location tracking, and radio frequency identification (RFID). These wireless sensor nodes will work efficiently for several years between battery replacements. This can be accomplished by the use of energy harvesting, utilizing ambient sources to prolong the life of the batteries in wireless sensor nodes.

Thin- film batteries are usually paired with a supercapacitor in order to handle the current surge when a wireless node transmits. As a result, supercapacitors are an unsubtle choice as energy buffers in energy harvesting applications. Unlike batteries, supercapacitors show extremely good cycle life and no issues relating to overcharge and over discharge. When the energy harvesting source is sufficient to meet the requirement of the wireless sensor node, then an adequately large supercapacitor may totally get rid of the need for a battery [1].

The energy harvesting system is made up of energy collection and energy storage. The collection part consists of the solar array. The energy storage device (supercapacitor), will as well serve as a power supply source to the SN. The sensor node in this project operates at RF 315MHz and is powered by a solar energy harvesting source. In order to properly power the sensor node, a supercapacitor of value 1.2F, and in most cases, operate at a voltage of 2.3 to 2.

The objective of this work is to save cost by independently powering wireless sensor nodes with an energy harvesting source without the use of disposable batteries that require constant replacement. In situation where the batteries are still in use by sensor nodes, a supercapacitor allows the sensor node to transmit its final data to the sink node, in the event of power failure, preventing data loss and its associated problem. Sometimes, the system in question is not properly studied. Also, it will be wise enough to gain more knowledge of how the energy harvested degrades with time through system modeling.

The paper is organised as follows: The introductory part of the paper provided in Section 1, deals with the general perspective and objective of the work. Section 2 reviews relevant work in energy harvesting, energy harvesting power sources and energy storage. Section 3 gives a detailed explanation of the system design and implementation, which includes design specification, methodology, modeling of the supercapacitor discharge and sizing of the supercapacitor. The result of the modeling and its discussion was presented in Section 4. The conclusion and future work that to be done was presented in Section 5.

2. REVIEW OF RELEVANT WORK IN ENERGY HARVESTING SYSTEM

Aaron et al. [3] worked on a wireless sensing platform utilizing ambient RF energy. They work on an ambient RF energy harvesting sensor node, which has onboard sensing and communication characteristics, and the system was developed and tested. In a similar development, Authors in [4], work on conventional MEMS generators, nanogenerators, and show that they have an added advantage of being flexible and foldable power sources, and mostly suited in the implantable biomedical sensors applications. Several work were done ranging from energy harvesting power sources [2,5,6]. In [2], it was shown that, the thermal energy harvesting is based on seebeck effect, and the available voltage is approximated by

$$V = \Delta S * \Delta T \tag{1}$$

where ΔS represents seebeck coefficients of each metal and ΔT beeing the temperature difference between them.

A summary of the amount of power available from each of the sources discussed above, which will help in in decision making and power budget can be viewed from [11].

2.1. Energy Storage

The free sources of energy for harvesting is not readily available. This poses as a disadvantage, therefore different methods of storing excess power to enable supply meet demand is very crucial. Energy is usually stored in a battery or a capacitor depending on its application. The second is used when the application needs to provide huge energy spikes while the first is used when devices needs to provide a steady flow of energy because it leaks less energy[5]. In order for thin film batteries to be capable of handling the current surge when a wireless node transmits; a large capacitor or supercapacitor is paired with them. Ultra- high density supercapacitors are preferable as energy buffers in energy harvesting [2].

Supercapacitors are used in this work as storage device to provide for long term back-up power to the wireless sensor nodes and help to get rid of battery back-up units, along with its maintenance and monitoring problems. It also helps to eliminate the environmental compliance issues that comes with battery disposal, and offers superior shelf life compared to batteries. A well sized supercapacitor will last for a long period of time.

2.1.1. Supercapacitor

Supercapacitors are also known as electric double- layer capacitor (EDLC). Ultracapacitor is the generic name used for a family of electrochemical capacitors. Capacitors are generally constructed by placing dielectric between opposed electrodes, which acts as capacitors by storing charges. Conventionally, energy is stored in capacitors when charge carriers (electrons) are removed from one metal plate and deposited on another. The potential between the two plates is created due to the charge separation, which can be utilized in an external circuit. Supercapacitors have a rare high energy density of several orders of magnitude greater than a high capacity electrolytic capacitor compared to common capacitors , though do not have a conventional solid dielectric. It utilizes the phenomena usually known as the electric double layer. The effective thickness of the "dielectric" is extremely thin in the double layer, and because of the porous nature of the carbon the surface area is exceedingly high, which means a very high capacitance.

The following are some of the advantages of supercapacitor as a storage device.

- They have high energy storage capacity compared to conventional capacitors;
- Their equivalent series resistance (ESR) is low compared to batteries, hence providing high power density;
- The can operate in low temperature up to $-40°C$ with minimal effect on efficiency;
- They have a simple charging method;
- They have high power density;
- Very fast charge and discharge;
- They are rugged and can be operated in harsh environmental condition due to their epoxy resin sealed case which is non-corrosive;
- Improve environmental safety;

- They have virtually unlimited cycle life.

The following are some of the disadvantages associated with supercapacitor as a storage device.

- They have a low voltage per cell, typically about 2.7V. In high voltage application, cells have to be connected in series;
- Due to their time constant, not suitable in Alternating and high frequency circuits.

Typically,they are used in applications where batteries have a shortfall when it comes to high power and life, and conventional capacitors cannot be used because of their inability to store enough energy. They offer a high power density along with adequate energy density for most short-term high power applications. A comparison of advantages and disadvantages of supercapacitors with other energy storage devices susc as batteries and the conventional capacitor can be found in a journal – Carbon materials for energy [9].

Figure 1. A real view of a supercapacitor

3. SYSTEM DESIGN AND IMPLEMENTATION

3.1 System Design Specifications

- CAP XX Supercapacitor: 1.2F, ESR of 50mΩ, rated voltage 5.5V, operating voltage (2.3 – 2.7) V cell.
- Sensor Node Voltage rating: (3 – 5) V.
- Power and current consumption of sensor node: 32mW and 8mA.
- Operating ambient temperature of Sensor node: (-20 to +85)^{0}C.
- Solar panel rating: Maximum Power Voltage = 5.82V, Maximum Power Current = 0.52A, Short circuit current = 0.55A, Open circuit voltage =7.38V, Output Power tolerance = 3%.

3.2. Design Methodology

This section gives an overview of the modules and operation of the entire system. The modules are as shown in Figure 2.

3.2.1. Operating Principle of the Energy harvesting wireless sensor nodes

The functional block diagram for the intended system is shown in Figure 3.1. The voltage source (Transducer) helps in converting the solar energy into an electrical signal (voltage). The voltage source is protected by an overvoltage circuit (shunt regulator). The third stage, the charging circuit, is a current-limiter circuit, and uses MOSFET to charge the supercapacitor in the final circuit (output circuit), which in turn store the energy used to charge the sensor battery or directly supply the sensor node. The overall system is subdivided into 4 basic modules.

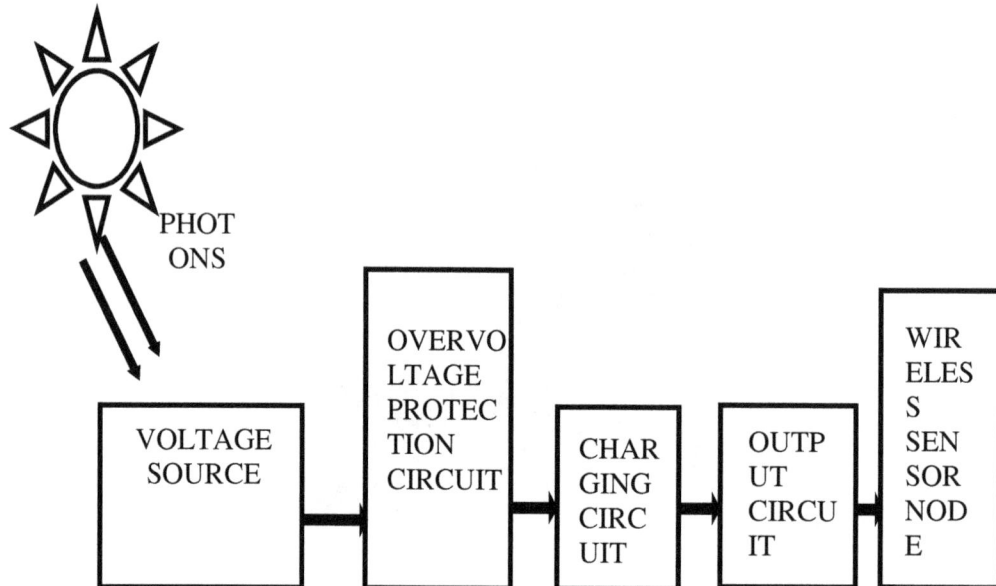

Figure 2. Block diagram of energy harvesting wireless sensor nodes

3.2.1.1. Voltage Source

Solar energy can be converted to electrical energy in the form of the voltage signal using solar cells or panel. The solar energy is as a result of illuminated junction of the cells. Its circuit model is shown in Figure 3. The current source, I_1, represents the current produced from electron-hole pair recombination due to solar radiation. The diode represents the solar cell's P-N junction characteristics. Current will pass through the solar cell just like it would pass througha diode when voltage is applied or produced across the terminals. The diode is characterized by its *ideality factor*, n, and its *reverse saturation current*. R_2 is the parallel resistance of the semiconductor materials, and diode current. R_1 is the series resistance of the metals used in the solar cell leads and contacts. Typically, $R_2 >> R_1$. The solar panel operating current and voltage is 0.52A and 5.82V respectively. It also has an open circuit voltage of 7.38V.

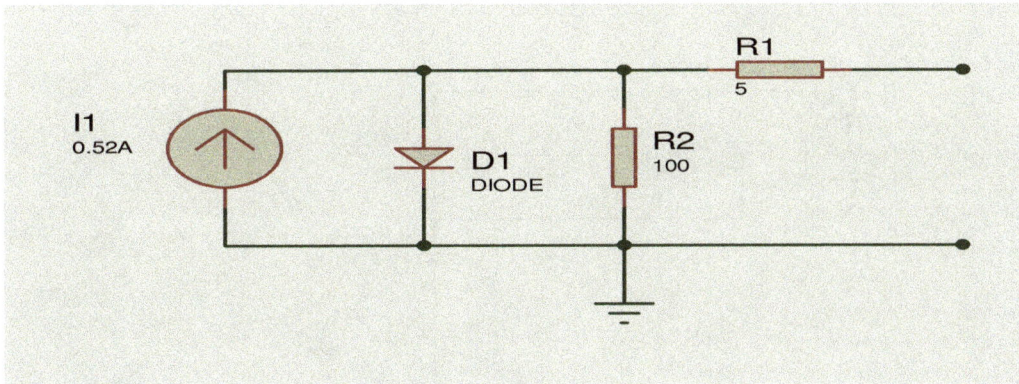

Figure 3.   Solar cell circuit model

3.2.1.2. Overvoltage Protection Circuit

Supercapacitors charges best by accumulating maximum current from the source of supply since they do not charge at a constant voltage like battery does. In our design specifications, the energy source's open-circuit voltage is greater than the supercapacitor's voltage, as such, we then requires overvoltage protection for our supercapacitor using a shunt regulator. It is usually used because of its simplicity and low cost. Once the supercapacitor is fully charged, it does not matter whether the excess energy dissipates. Figure 3.3 shows the overvoltage protection circuit.

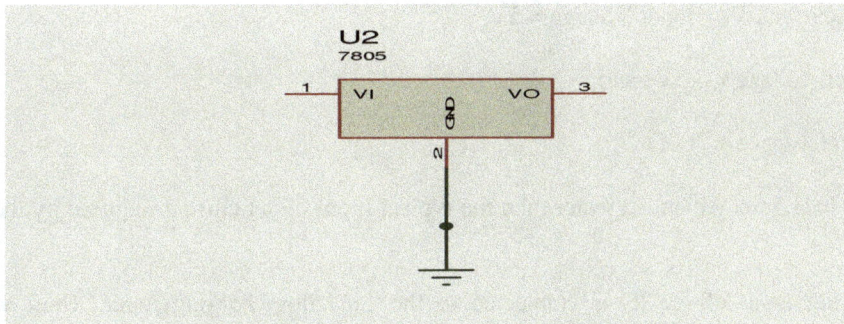

Figure 4.   Fixed output voltage regulator

The voltage regulator used is LM7805, a 5V, 1A regulator. Since its minimum input voltage is 7V, it is suitable to function properly with the 7.38V delivered by the solar panel and it will maintain an output voltage of 5V.

3.2.1.3. Charging Circuit

This is basically switching/current limiting stage. We will be using a MOSFET for this purpose. In this case, we shall be using MOSFET in load switching application. In this module, we employed CAP-XX supercapacitor with high capacitance and low ESR on the power rail, which will now serves as the load that the MOSFET will be seeing. The supercapacitor is re-charged between load pulses. We will be using MOSFET to charge the supercapacitor in a current-limiting mode.

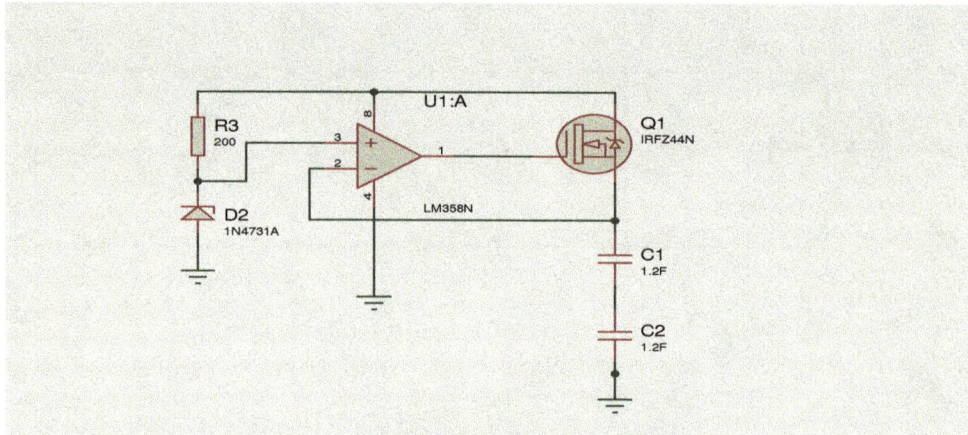

Figure 5.   Simple solar cell charging circuit

The charge controller is built around the IC LM358 low power dual operational amplifier configured as a comparator. A 4.3V zener diode is connected to the non-inverting input (pin 3) of the IC leading to a constant 4.3V at this input pin. The zener diode current available at this pin is obtained as

$$I = \frac{V_{in} - V_{out}}{R} \qquad (2)$$

Where, I = current, V_{in}= input Voltage = 5V,

V_{out} = output voltage = 4.3V   and

R = Series resistance = 200Ω

Therefore, I= 3.5mA which is greater than the typical input offset current required by the IC (i.e. 2nA).

The inverting input of the IC is connected to the capacitor charging input. Thus, when the capacitor is charging, the IC compares the capacitor charge voltage with the 4.3V in pin 3. If the charge voltage is less than 4.3V, the output of the comparator (pin 1) remains at a voltage of   Vcc - 1.5V (LM358 datasheet)

which is obtained as

V_{out} = V_{cc} - 1.5V = 5 - 1.5 = 3.5V

This voltage serves as the gate voltage of the N-channel MOSFET IRFZ44N with ratings:

Gate to Source Voltage V_{gs} = 10V at 25°C

Continuous Drain Current I_D = 50mA at 25°C

Minimum Gate Threshold Voltage $V_{gs\,(th)}$ = 2V

From the trans inductance curve of the MOSFET,

$$I_D = K \ (V_{gs} - V_{gs\,(th)})^2 \tag{3}$$

$$K = \frac{I_D}{(V_{gs} - V_{gs\,(th)})^2}$$

$$K = 780 \ (mA/V^2)$$

K is a constant, thus Drain current I_D, for Vgs = 3.5V is obtained as

$$I_D = 0.78 \ X \ (3.5 - 2)^2 \ = \ 1.755A$$

This means that the MOSFET can conduct up to 1.755A with the input voltage of 3.5V and this value is ok since only 0.52A flows across the drain current.

3.2.1.4. Output Circuit

The output circuit basically consists of the supercapacitor that feeds the wireless sensor node. The supercapacitor is to power a sensor node operating at (3 – 5) V. The temperature at which the nodes operate is also between -20 ^{0}C to +85 ^{0}C. An ideal supercapacitor circuit model is shown in Figure 6. The supercapacitor should be sized, because, supercapacitor cells typically operate at 2.3 to 2.7V.

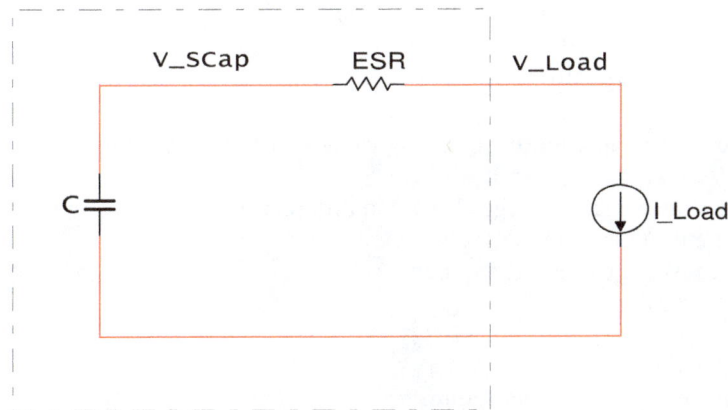

Figure 6.   An ideal supercapacitor circuit model

3.3. Modeling the Supercapacitor Discharge

Supercapacitor cells typically operate at 2.3 to 2.7V. Limiting its charge voltage to less than the cell-rated voltage and storing enough energy for a particular application is most efficient and cost effective method. An easier approach method to sizing it is to calculate the energy necessary to support the peak power of the application, P.t, and set this value equal to $0.5 \times C \times (V_{initial}^2 - V_{final}^2)$, where C is the capacitance, $V_{initial}^2$ is the square of the supercapacitor's voltage just before the peak power burst, and V_{final}^2 is the square of the final voltage. But this equation does not allow for any losses in the supercapacitor's ESR (equivalent series resistance). The load sees a voltage of $V_{initial} - ESR \times I_{Load}$, where I_{Load} is the load current. Due to the decrease in load voltage, the load current increases to achieve the load power. Referring to Figure 6, our supercapacitor discharge can be modelled as:

$$V_{Load} = V_{Scap} - I_{Load} \times ESR; \qquad (4)$$

$$P_{Load} = V_{Load} \times I_{Load} = \left(V_{Scap} - I_{Load} \times ESR\right) \times I_{Load}$$

$$= V_{Scap} \times I_{Load} - I^2{}_{Load} \times ESR \qquad (5)$$

where V_{scap} is the supercapacitor's voltage

The equation further leads to load current, i.e.

$$I^2{}_{Load} \times ESR - V_{Scap} \times I_{Load} + P_{Load} = 0 \qquad (6)$$

Supercapacitor discharge can then be simply modelled using Matlab as:

$$I_{Load}(t) = \left[\frac{V_{Scap}(t) \pm \sqrt{(V_{Scap}(t))^2 - 4 \times ESR \times P_{Load}}}{2 \times ESR} \right] \qquad (7)$$

$$V_{Load} = V_{Scap}(t) - I_{Load} \times ESR \qquad (8)$$

$$V_{Scap}(t + \Delta t) = V_{Scap}(t) - \frac{I_{Load}(t) \times \Delta t}{C} \qquad (9)$$

3.4. Sizing the Supercapacitor

In order to size the supercapacitor the following variables need to be defined.

    I.   Maximum charged voltage (Vmax), if different from the working voltage (Vw).
    II.   Minimum Voltage (Vmin)
    III.   Power (W) or current (I) required.
    IV.   Discharge duration (t)
    V.   Duty cycle.
    VI.   Required life
    VII.   Average Operating temperature.

The last three (3) variables are used to calculate the life degradation factor to use for the supercapacitor. This is not part of the project scope. Determination of the appropriate size and also number of cells required for our application we proceed as follows:

During the discharge cycle of a supercapacitor there are two variables to look at. The voltage drop due to equivalent series resistance (ESR) and the capacitance, [10].

From [10]; Voltage drop due to the equivalent series resistance is obtained as

$$V_{ESR} = I_{Load} \times ES \qquad (10)$$

Voltage drop due to the capacitance of the supercapacitor

$$V_{Scap}(t) = \frac{I_{Load}(t) \times \Delta t}{C} \qquad (11)$$

The total voltage drop is therefore obtained as:

$$V_{dt} = I_{Load} \times ESR + \frac{I_{Load}(t) \times \Delta t}{C} \tag{12}$$

V_{dt} is the total voltage drop when the capacitor is discharged. This is equal to the difference of Vw (working voltage) and Vmin (minimum voltage) as shown in Figure 3.6. Allowing a larger V_{dt} will reduce the capacitance size used. Usually by allowing the capacitor to drop to 0.5 Vw, 75% of the capacitor energy is discharged.

I_{Load} = current used to discharge the supercapacitor in amperes. For equation (12) we assume

I_{Load} to be a constant current discharge.

Δt = time taken to discharge the capacitor between Vw and Vmin in seconds.

C = Total capacitance of the supercapacitor in farad. If a single cell is used, then it is referred to as the cell capacitance. The equation 13 below shows the equivalent capacitance gotten from the number of capacitors in series or parallel when more than one cell is used.

$$C = Cc \times \frac{Number\ of\ capacitors\ in\ parallel}{Number\ of\ capacitors\ in\ series} \tag{13}$$

C_C = cell capacitance.

ESR = total resistance of the supercapacitor in ohms. If a single cell is used, then it is called the cell resistance. The equation 14 below also shows the equivalent resistance based on the same condition as equation 13.

$$ESR = ESRcell \times \frac{Number\ of\ resistors\ in\ series}{Number\ of\ resistors\ in\ parallel} \tag{14}$$

From the system design specification;

V_{max} = 5V
V_{min} = 3V
I = 0.008 A
P = 0.032W

Using the values above let us determine the size of the supercapacitor.

To determine the value of our stack supercapacitor, recall that, Energy is given as

$$P \times \Delta t = 0.5 \times C \times (V_{initial}^2 - V_{final}^2)$$

Therefore, $C = \frac{2 \times P \times \Delta t}{V_{initial}^2 - V_{final}^2} = \frac{2 \times 0.032 \times 150}{16} = 0.6F$

But, $ESR = ESRcell \times \frac{Number\ of\ resistors\ in\ series}{Number\ of\ resistors\ in\ parallel}$

Therefore, $ESR = 0.05 \times 2 = 0.1\ ohms$

From equation (12), neglecting the effect of the ESR,

$$V_{dt} = \frac{I_{Load}(t) \times \Delta t}{C} \text{ , therefore } \quad \Delta t = \frac{V_{dt} \times C}{I_{Load}(t)}$$

where $V_{dt} = 5 - 3 = 2V$

$$\Delta t = \frac{2 \times 0.6}{0.008)} = 150 \ seconds = 2.5 \ minutes.$$

Given Vmax = 5V,

since each supercapacitor cell is usually rated at 2.7V, then we divide Vmax by 2.7 and round up.

$\frac{5}{2.7} = 1.85 = 2 \ approximated$, 2 cells in series are required.

From equation (13), the cell needed will be in the range of

$$C = Cc \times \frac{Number \ of \ capacitors \ in \ parallel}{Number \ of \ capacitors \ in \ series}$$

$$C = Cc \times \frac{Number \ of \ capacitors \ in \ parallel}{Number \ of \ capacitors \ in \ series} = \frac{1.2 \times 1}{2} = 0.6F$$

Each of the 1.2F cells will have a rated voltage of 2.7V, since they will be connected in series, hence the total voltage will be 5.4V which will be enough to drive our wireless sensor node. Considering the effect of ESR, from equation (12), i.e

$$V_{dt} = I_{Load} \times ESR + \frac{I_{Load}(t) \times \Delta t}{C},$$

$$\Delta t = \frac{(V_{dt} - I_{Load}(t) \times ESR) \times C}{I_{Load}(t)} = \frac{(2 - 0.008 \times 0.1) \times 0.6}{0.008} = 150 \ seconds = 2.5 \ minutes.$$

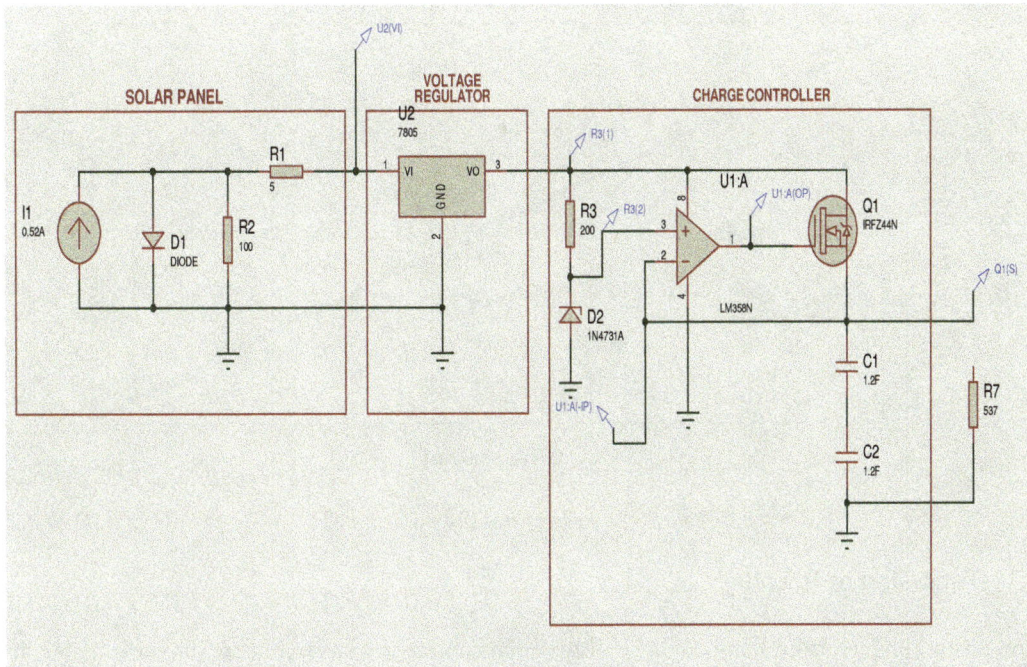

Figure 7.   Overall circuit diagram of energy harvesting wireless sensor nodes

4. Results from Modeling

Table 1.Result of the Modeling of Supercapacitor discharge voltage

Discharge Voltage (V)	Discharge duration of the capacitor (minutes)
5.4000	0
4.8073	1
4.2147	2
3.6220	3
3.0294	4

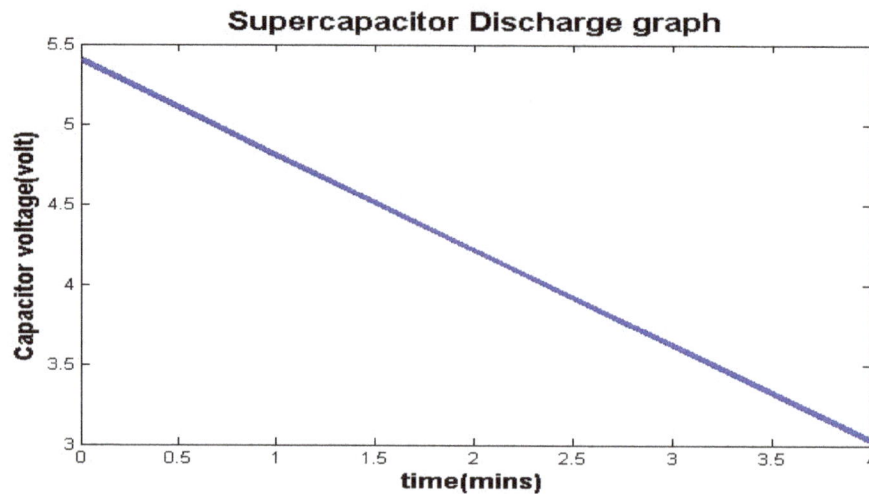

Figure 8.Supercapacitor discharge graph

4.1. Discussion of Results

The result obtained from the modelling of the supercapacitor discharge is as shown in Table 4.0. In the table, the maximum value the capacitor is charged to is 5.4V and the minimum operating voltage which the wireless sensor node that was backed up by the capacitor can tolerate before it stops working is 3.0294V. The specification for the wireless sensor node requires a maximum voltage of 5V and a minimum voltage of 3V. This implies that , only a part of the stored energy is available for applications, since the voltage drop and the time constant over the internal resistance mean that some of the charge stored cannot be accessed. The voltage drop is used in determining the size of the supercapacitor capacitance to choose for a given application. A larger drop in voltage when the capacitor is discharged will reduce the capacitance size used. When the voltage drop is more than the allowable voltage drop, the chosen capacitance is sized up and vice versa. Typically by allowing the capacitor to drop to half of the working voltage of the supercapacitor, 75% of the capacitor energy is discharged.

5.0 CONCLUSION AND FUTURE WORK

In conclusion, it is evident that some form of energy harvesting either from a single or multiple sources is needed for low power wireless sensor nodes- so as to reduce maintenance and prolong the lifecycle of the devices. Energy harvesting using solar cell is an interesting alternative source of energy that has the potential to provide energy independence to wireless sensor devices without the use of expensive wires, or batteries that will need replacement every now and then. From our results, it is can be said that, only a part of the stored energyis available for applications, we can now conclude that a larger drop in voltage when the capacitor is discharged will reduce the capacitance size used and this knowledge can be used to further strengthened simulation approach and practical implementation of energy harvesting wireless sensor network.

In general, there is need for optimal design considerations and energy budget for a successful implementation in any energy harvesting system. Reducing the operation of the energy of wireless sensor nodes will give room for a smaller storage capacitor value to be used, and by that, the cost of the system is reduced and sensitivity and efficiency due to the lower parasitic leakage of smaller capacitors is improved. There is a need for more than one energy harvester for a successful energy harvesting application.

REFERENCES

[1]   Stojcev, M.K., Kosanovic, M.R., and Golubovic, L.R. Power management and energy harvesting technique for wireless sensor nodes. Proceeding of the 9th International Conference on Telecommunication in Modern Satellite, Cable, and Broadcasting Services, 2009: 65-72.

[2]   John Donovan, Energy Harvesting for Lower- Power Wireless Sensor Nodes, Contributed By Convergence Promotions LLC, 2012 [online].   Available from http://www.digikey.com/en-US/articles/techzone/2012/feb/energy-harvesting-for-low-power-wireless-sensor-nodes/ [Accessed 16th April, 2014].

[3]   Aaron N. Parks, Alanson P. Sample, Yi Zhao, Joshua R. Smith. (2013). "A Wireless Sensing Platform Utilizing Ambient RF Energy" [Online].   Available from sensor.cs.washington.edu/......./wisnet_.........[Accessed 17th April, 2014].

[4]   Action Nechibvute, Albert Chawanda, and Pearson Luhanga, (2012), "Piezoelectric Energy Harvesting Devices: An Alternative Energy Source for Wireless Sensors", Smart Materials Research, Volume 2012 (2012), Article ID 853481, 13 pages [online].   Available from http://dx.doi.org/10.1155/2012/853481 [Accessed 20th April, 2014].

[5]   Wikipedia.   (2014).   Energy   harvesting   [online].   Available   from http://en.wikipedia.org/wiki/Energy_harvesting/ [Accessed 6th April, 2014].

[6]   [online] available at http://en.wikipedia.org/wiki/Thermoelectrics#cite_note-14

[7]   Wikipedia.   (2014).   Supercapacitor   [online].   Available   from http://en.wikipedia.org/wiki/Supercapacitor/ [Accessed 6th April, 2014].

[8]   http://www.tecategroup.com/ultracapacitors-supercapacitors/ultracapacitors-FAQ.php#What_is_an_ultracapacitor

[9]   Pierre Mars (2012). "Coupling a Supercapacitor with a Small Energy Harvesting Source",Published in EE Times Design [online].   Available from http://www.cap-xx.com [Accessed 18th April, 2014].

[10]   Maxwell Technology Applications Note: Ultracapacitor Cell.

[11]   Adamu Murtala Zungeru, Li-Minn Ang, SRS. Prabaharan, Kah Phooi Seng (2012), "Radio Frequency Energy Harvesting and Management for Wireless Sensor Networks", Publisher: CRC Press, Taylor and Francis Group, USA, [online], Available: http://www.crcnetbase.com/doi/abs/10.1201/b10081-16.

13

CCABC: Cyclic Cellular Automata Based Clustering For Energy Conservation in Sensor Networks

Indrajit Banerjee[#], Prasenjit Chanak[*], Hafizur Rahaman[#]

[#]Department of Information Technology
[*]Purabi Das School of Information Technology
Bengal Engineering and Science University, Shibpur, Howrah, India.

[1]ibanerjee@it.becs.ac.in, [2]prasenjit.chanak@gmail.com,
[3]rahaman_h@it.becs.ac.in

Abstract—*Sensor network has been recognized as the most significant technology for next century. Despites of its potential application, wireless sensor network encounters resource restriction such as low power, reduced bandwidth and specially limited power sources. This work proposes an efficient technique for the conservation of energy in a wireless sensor network (WSN) by forming an effective cluster of the network nodes distributed over a wide range of geographical area. The clustering scheme is developed around a specified class of cellular automata (CA) referred to as the modified cyclic cellular automata (mCCA). It sets a number of nodes in stand-by mode at an instance of time without compromising the area of network coverage and thereby conserves the battery power. The proposed scheme also determines an effective cluster size where the inter-cluster and intra-cluster communication cost is minimum. The simulation results establish that the cyclic cellular automata based clustering for energy conservation in sensor networks (CCABC) is more reliable than the existing schemes where clustering and CA based energy saving technique is used.*

Keywords— *Modified cyclic cellular automata (mCCA), clustering, wireless sensor network (WSN), base station (BS)*

I. INTRODUCTION

Wireless sensor network (WSN), consisting of thousands of wireless sensors, is bounded due to the limited computational capability, battery power and memory capability of its components. The sensor nodes are deployed in a monitoring area and communicate among themselves following the multi-hop wireless communication. The information received at a node is computed and communicated to the nearest base station [1]. In homogeneous networks, all the sensor nodes are identical in terms of battery energy/power and hardware complexity. The energy saving network design is the major issue in WSN to increase the life time of network nodes.

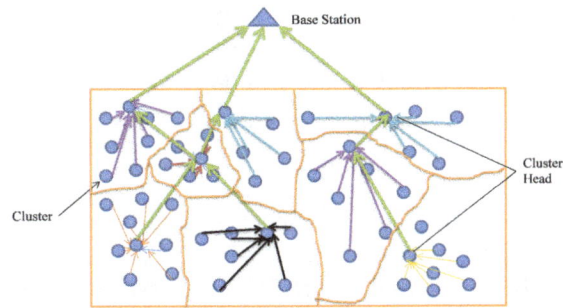

Fig.1: Cluster base communication in wireless sensor network

In WSN, a base station (BS) is stationary and the sensor nodes may be movable [2]. The energy loss in a node is very high when a node directly communicates with the base station (Fig.1). On the other hand, in a cluster based node management scheme, a node communicates with the BS through a leader, called cluster head (CH) (Fig. 1). In the clustered scheme proposed so far [2], [3] and [4], each and every node is in active state, therefore, a particular area is monitored by the two or more nodes. The schemes LEACH [2], EEPSC [3], LEACH-C [4], UCCP [5], EECS [6], EEDUC [7] and DDC [8] cannot protect the network nodes from early energy dissipation lead to short span of life. As a node expires within a short time, the new sets of clusters are formed very frequently. This demands massive message exchanges among nodes and, therefore, causes the uncontrolled power dissipation. The cellular automata based techniques [1], [9] and [10] proposed so far recover the common region sensing problem, which ensure that the number of nodes in active state is minimum. The active nodes communicate directly with the base station leading to unwanted energy loss in the network.

In this context, the CCABC technique, proposed in this work, develops a cluster based network management system that ensures full coverage of the sensor network with minimum number of active node. If any active nodes fail to sense, transmit or receive data then this node is declared as a dead node and a neighbouring stand-by node is selected through CCABC scheme for efficient replacement. In the CCABC scheme, the clusters generated are of optimal size, in which the data are aggregated properly for further reduction of overhead in data processing. The member nodes of that optimal cluster size can send their data to the cluster head with minimum energy loss (Fig.2). The proposed scheme also determines the position of the cluster head where each node of the cluster sends their data with minimum energy loss. The CCABC selects a cluster head from the nodes in an energy efficient manner.

The organization of this paper is as follows. Modified cyclic cellular automata (mCCA) are elucidated in Section 2. The mathematical model of the proposed scheme is described in Section 3. In Section 4, we have introduced the proposed algorithms, developed over mCCA. The simulation results are reported in Section 5. Finally we conclude our paper in Section 6.

II. CYCLIC CELLULAR AUTOMATA

The cyclic cellular automata (CCA) follow a local rule which is same for all states S. Each cell in CCA contents different states from state range S= {0, 1, 2...k-1} [11], [12], [13] and [14]. The integer k is the maximum number of state. In the exiting field Z^2 all cells change their states within S.

$$I_t(P) = Z^2 \rightarrow \{0,1,2,...,k-1\} \tag{1}$$

The CCA generates a spiral structure when cells are changing their states from zero to k-1 as equation number 1. The $I_t(P)$ represents the present state of a cell P ϵ Z^2 at integer time t. If and only if for some given threshold value θ at *Von Neumann neighbour* set N(x) a cell P changes its state $I_t(P)$ to another state $I_{t+1}(P)$ at time t+1 is shown in equation 2.

$$I_{t+1}(P) = I_t(P) + 1 \bmod k \qquad (2)$$

The threshold value θ represents the y number of neighboring cell's condition in set N(x) within the field, y ϵ N(x). The neighboring cells set N(x) are also in k-1 state. The initial state of automaton is said to be primordial soup [11]. A classical model of excitable media was introduced in 1978 by Greenberg and Hastings (GH) [13] and [14] described next.

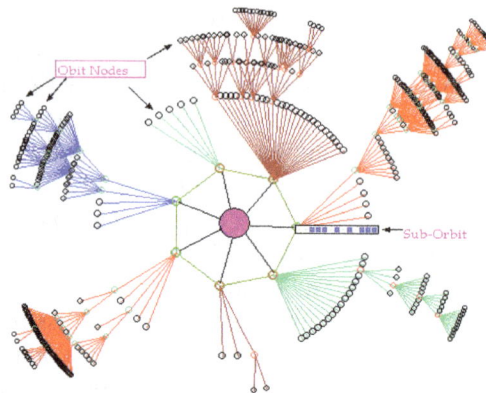

Fig.2: Cluster head connected with cluster nodes within a cluster.

Greenberg-Hastings model (GHM)

The Greenberg – Hastings model [12] is a simplified cellular automaton that is run in excitable medium. In GHM according to state change rule each cells of the automata changes their state and produced a special type node pattern. The state change rules of the CCA in GHM are described bellow

1. If $\gamma_t(x)$ = n, then $\gamma_{t+1}(x = n + 1 \bmod e\ k)$

2. If $\gamma_t(x)$ = 0 and at least n neighbours are in state 1 then $\gamma_{t+1}(x)$ = 1; otherwise the current state (0) is continued.

Where $\gamma_t(x)$ is cell's condition (or state) at time t and $\gamma_{t+1}(x)$ is the next state of cell at time t+1. In our proposed modified cyclic cellular automata based EERIH [19] we have modified the GSM rules. The EERIH generate special type of nodes pattern and arrange the active nodes into some cluster. This pattern also helps us for routing the data from cluster head to base station in energy efficient manner.

Proposed modified CCA

In our modified cyclic cellular automata (mCCA) based scheme every cell changes it state according to the nine neighbours' cells state condition. The state change rules of the cells are defined below [19]:

1. If cell's present state is $\delta_t(p) = n$. where $n > 0$ *and* $n < I - 1$, then next state of the cells is $\delta_{t+1}(p) = n + 1$

2. If cell's present state is $\delta_t(p) = n$. where $n = I - 1$, then next state of the cells is $\delta_{t+1}(p) = 0$.

3. If cell's present state is $\delta_t(p) = 0$, then they check their neighbour cells state and if θ numbers (threshold value) of nodes are present in nonzero state then next state is $\delta_{t+1}(p) = 1$ otherwise they are not changing their state i.e. $\delta_{t+1}(p) = 0$.

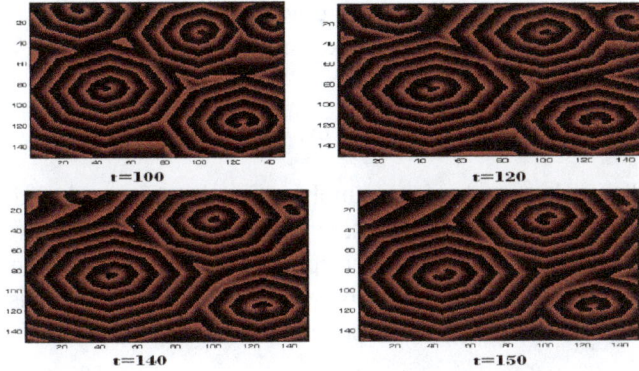

Fig.3: Atom ic structure of node pattern generated by CCA

Where $\delta_t(p)$ is the state of the cells at time t and $\delta_{t+1}(p)$ is the state of cells at t+1 time. Number of state is {0, 1...I-1}. With the help of this state change rule we are arranging every sensor node in a pattern of atomic structure. In this scheme every node is changing their state in a time interval and nodes pattern are controlled from the primordial soup. The nodes pattern of the proposed cyclic cellular automata shows in Fig.3 where with time every node is changing their state i.e. spirals are propagated.

III. MATHEMATICAL ANALYSIS

In this section we have analysed the mathematical model of our proposed scheme energy efficient clustering scheme for wireless sensor networks. Here we calculate effective cluster size where each node sends their data with minimum energy loss. In energy efficient clustering scheme for wireless sensor networks, two types of data communication can take place among nodes in the network. These are intra-cluster and inter-cluster data communication. The inter-cluster data communication is the data transfer between cluster head and base station as shown in Fig. 1. In the proposed scheme the active nodes are self organized into an atomic structure to form clusters (see Section IV). The intra-cluster communication cost is the energy spent by all orbital nodes to send their data to the cluster head. It involves intra-orbit and inter-orbital data communication cost. In intra-orbital data communication, the sub-orbital node transmits their data to the nucleus of the orbit (Fig.2). In inter-orbital data communication, orbits are sending their data to the nucleus of nearest upper layer orbit. If we consider the total sensor network as a single cluster, then inter-cluster communication cost is zero but the intra-cluster communication cost is very high. On the other hand if cluster size is zero, then intra-cluster communication cost is zero but in this case inter-cluster communication cost is very high. With the help of inter-cluster communication cost and intra-cluster communication cost we can formulate an efficient cluster size in cluster based energy efficient wireless sensor networks management scheme, where each node sends their data with minimum energy loss. We have find out the position of the cluster head within a cluster where the communication energy loss is minimal. In wireless sensor network different sensor nodes send their data to base station from different locations. Hence the transmission range of every sensor node is different as well as energy loss of the sensor nodes is also different. In the proposed model every sensor node sends their data with minimum energy loss. These are calculated by energy loss errors in the sensor network. We are also introducing an efficient data aggregation formula of the proposed technique.

Definition 1: Let X be the monitoring field, covered by the set of sensor node and $f(\emptyset)$ is the inter-cluster communication cost that is depending on transmission distance and nodes density. Therefore,

$$f(\emptyset) = \int_{e_i \in M_c} \mu \times \beta_s^c ((\varepsilon + \omega) + \gamma d_i^n) dd_i \qquad (3)$$

The size of data is β_s^c, in cluster c having s number of nodes. The nodes' density is μ. The ε is the energy consumed in the transmitter circuit, ω is the dissipated energy for data aggregation and γ is the dissipated energy in the transmitter op-amp. The e_i is the cluster head that collects all data of a single cluster and transmits it to base station. Transmission distance between two cluster head e_i is d_i and n is a path-loss exponent. M_c is the number of member node in a cluster.

Definition 2: The intra-cluster communication cost of the node is f(I) that depends on the cluster size and transmission distance between cluster head and cluster member nodes. Therefore,

$$f(I) = \int_{e_i \in P_c} \alpha_i (\varepsilon + \gamma d_j^n) dd_j \qquad (4)$$

In above equation α_i is the total bits transmitted along edge e_i. P_c is number of nodes in cluster, which are used to collect data from each orbit in a cluster c. The distance between the transmitter and receiver node in a cluster is d_j.

Theorem 1: Inter-cluster communication cost $f(\emptyset)$ is single-valued and possesses a unique derivation with respect to $\emptyset$ at all points of a WSN region R (in WSN each nodes are connected by multi-hop communication) is called an analytic or a regular function of $\emptyset$ in that region. A point at which an analytic function ceases to possess a derivation is called a singular point of the function.

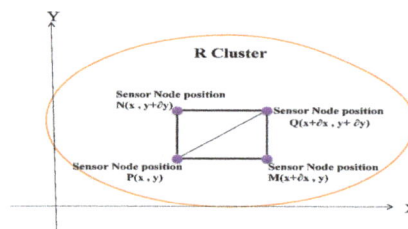

Fig.4: Close Region R in WSN, where P and Q are any node position

Proof- Let $\omega = f(\emptyset)$ be a single-valued function in a region within the WSN of the variable $\emptyset = x + y$ (Fig. 4). Then the derivative of $\omega = f(\emptyset)$ is defined the inter-cluster communication cost for data transmission between the cluster head,

$$\frac{d\omega}{d\emptyset} = f'(\emptyset) = \underset{\partial\emptyset \to 0}{\text{Lt}} \frac{f(\emptyset + \partial\emptyset) - f(\emptyset)}{\partial\emptyset} \tag{5}$$

provided the limit exists and has the same value for all the different ways in which $\partial\emptyset$ approaches to zero. In sensor network every node is virtually connected to each other, each orbital is virtually connected to upper orbital and cluster heads transmit data to base station with the help of other cluster heads (Fig. 2). Suppose $P(\emptyset)$ is fixed node position within a region R and $Q(\emptyset + \partial\emptyset)$ is a neighbouring node position (Fig. 4). The node Q may approaches towards P along any straight or curved path in the given region R, i.e. $\partial\emptyset$ may tend to zero in any manner and $\frac{d\omega}{d\emptyset}$ to exist.

Theorem 2: The inter-cluster communication cost $f(\emptyset)$ is analytic in the cluster region D between two simple close clusters X and X1, then

$$\int_X f(\emptyset)d\emptyset = \int_{X1} f(\emptyset)d\emptyset \tag{6}$$

Where X represents the whole sensor network as a cluster and the X1 is the other cluster within the cluster X.

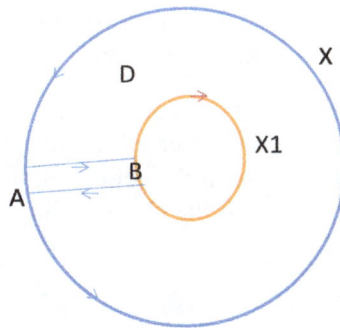

Fig.5: X1 is a close cluster under a close cluster X.

Proof- We introduce the cross-cut AB in region X. Then $\int f(\emptyset)d\emptyset = 0$. Where the path is as indicated by arrows in Fig. 5; i.e. along AB, X1 in clockwise sense & along BA, X in anti-clockwise sense. Therefore, $\int_{AB} f(\emptyset)d\emptyset + \int_{X1} f(\emptyset)d\emptyset + \int_{BA} f(\emptyset)d\emptyset + \int_X f(\emptyset)d\emptyset = 0$ But, since the integration along AB and BA cancel each other, it follows that $\int_X f(\emptyset)d\emptyset + \int_{X1} f(\emptyset)d\emptyset = 0$. Reversing the direction of the integral around X1 and transposing, we get $\int_X f(\emptyset)d\emptyset = \int_{X1} f(\emptyset)d\emptyset$ here each integration being taken in the anti-clockwise sense. If X1, X2, X3 ...... be any number of close cluster within close cluster X then,

$$\int_X f(\emptyset)d\emptyset = \int_{X1} f(\emptyset)d\emptyset + \int_{X2} f(\emptyset)d\emptyset + \int_{X3} f(\emptyset)d\emptyset \tag{7}$$

Theorem 2 proves that the total inter-cluster communication cost $f(\emptyset)$ of any number of internal clusters are same as that of whole network.

Theorem 3: The inter-cluster communication cost $f(\emptyset)$ is analytic within a close cluster and if point a is any node position within X, then energy loss at that node is

$$f(a) = \frac{1}{2\pi} \int_X \frac{f(\emptyset)}{\emptyset - a} \, d\emptyset \tag{8}$$

Proof- Let us consider the function $\frac{f(\emptyset)}{(\emptyset - a)}$ which is analytic at all nodes position within X except at $\emptyset = a$ node position with the node position a as centre of cluster and r is the radius of cluster area. We draw a small circle cluster lying entirely within X. Now $f(\emptyset)/(\emptyset - a)$ being analytic in the region enclosed by X and X1. $\int_X \frac{f(\emptyset)}{\emptyset - a} d\emptyset = \int_{X_1} \frac{f(\emptyset)}{\emptyset - a} d\emptyset$ $= \int_{X_1} \frac{f(a + re^\theta)}{re^\theta} re^\theta d\theta = \int_{X_1} f(a + re^\theta) d\theta$ for any nodes on the network, $\emptyset = a + re^\theta$ and $d\emptyset = re^\theta d\theta$. In the limiting form, as the circle cluster X1 shrinks to the node position a, as $r \to 0$ we consider every sensor node as a point. The integral approaches to $\int_{X_1} f(a) d\theta = f(a) \int_0^{2\pi} d\theta = 2\pi f(a)$ $f(a) = \frac{1}{2\pi} \int_X \frac{f(\emptyset)}{\emptyset - a} d\emptyset$ In general, $f^n(a) = \frac{n!}{2\pi} \int_X \frac{f(\emptyset)}{(\emptyset - a)^{n+1}} d\emptyset$ X is a large monitoring area, and data is travelling among the nodes in sensor network then $|\emptyset - a| = r$. In this reason $\emptyset$ is unevenly distributed within the close cluster X.

$$|f^n(a)| \le \frac{Mn!}{r^n} \tag{9}$$

Where M is the maximum value of $|f(\emptyset)|$ on cluster X.

In-order to find out the actual position of the cluster head in the network, we have divided the whole network into equal size of clusters. The suitable position of the cluster head can be determined according to the following theorem.

Theorem 4: The analytic function inter-cluster communication cost f($\emptyset$) within the close cluster region is average at the centre position of a cluster region. In this region f(I) is very small. The centre position, where inter-cluster communication cost f($\emptyset$) is average and intra-cluster communication cost f(I) is small, is the cluster head location.

Proof- When we consider the cluster of sensor node as a circle then inter-cluster communication cost of the cluster node $f(\emptyset)$ is average in the centre point of the cluster. Because inter-cluster communication cost $f(\emptyset)$ depends mainly on the distance between cluster head and base station. On the other hand intra-cluster communication cost $f(I)$ also depends on the member nodes to cluster head distance. If we select cluster head as a nearest node of the base station then the inter-cluster communication cost is minimum but within the cluster member nodes and cluster head, distance is increased (so the value of f(I) is high), therefore, total energy loss by the cluster f($\emptyset$) + f(I) is increased. Within cluster long distance cluster member nodes lose more energy. So we are going to select cluster head's position at the median of the cluster.

Theorem 1 describes inter-cluster communication cost f($\emptyset$), which is an analytic function, because this function follows the necessary and sufficient condition of an analytic function. We are selecting some nodes position of whole network according to Theorem 2 and start the CCA spiral propagation for the selection of nodes' position. With time the CCA spiral propagation cover the entire network and therefore f($\emptyset$) value is increased and f(I) value is decreased. When the inter-cluster communication cost and intra-cluster communication cost is equal i.e. f($\emptyset$) = f(I), the spiral propagation (see Section IV) will stop at that point. This is the optimal size of the cluster where data are aggregated properly according to the equation no. 11. The member nodes of that optimal cluster size can send their data with minimum energy loss. With the help of Theorem 3 we are going to calculate inter-cluster communication cost in each sensor node and determine the position of the nodes from where CCA spiral start to propagate. Theorem 4 determines the position of the cluster head where each node of the cluster sends their data with minimum energy loss.

Data Aggregation Model

Data aggregation on a sensor network depends on data correlation. With the help of data correlation we can aggregate efficient amount of data from the nodes. The aggregated data is then transmitted to cluster head (CH). A large amount of energy is wasting due to transmission of same type of information to the base station. When density of the active nodes increases to provide fault tolerant feature in the sensor network [16], large number of number of nodes cover a particular area which sense similar information. If density of the nodes is very high, then a large amount of energy is wasted for transformation of same information through multi-hop sensor network. Different types of approaches have been proposed to model the correlation of data; one of them, entropy-base model is very popular. The entropy-base data correlation and compression algorithm is described in [17].

$$B_s(d_0) = b_0 + (s-1)(1 - \frac{1}{\frac{d_0}{c}+1})b_0 \tag{10}$$

Here d_0 is the inter-node distance and b_0 is the number of bits generated by each source, the constant parameter c characterizes the spatial data correlation. $B_s(d_0)$ is the number of compressed bit messages generated by the cluster head in a s-node cluster. The entropy-base model is applied in the CCABC scheme, as the model aggregates data accurately and efficiently. In CCABC, we have modified the equation no. 18 in-order to decrease the data aggregation error rate.

$$B_s(d_0) = b_0 + (s-1)(1 - \ln 2^{e^{-\frac{d_0}{\sigma+s}}})b_0 \tag{11}$$

σ represents the minimum size of the cluster. Other parameters have the same meaning as in equation no 18. The correlation error reduces in our proposed equation. If the cluster size increases the data correlation error increases. So we consider here the optimal cluster size.

IV. CCABC ALGORITHM

The modified CCA (see Section II) is used to create clusters in sensor network. During the change of state the spirals are decomposed randomly in a time interval and generate wave pattern. In this way nodes are self-organized into an atomic structure (AS). In an atomic structure the nodes are distributed into orbits and nucleus. The orbit is also divided into sub-orbits. The proposed cluster formation algorithm is as follows:

CCABC (Algorithm 1): Cluster Generation

Insert all nodes into S (array of nodes)

CS (Cluster Set) = Null (empty set)

WHILE S! = Null **DO**

    Calculate Nodes energy from Algorithm 3

Set CC = Si (single Cluster)

Set Inter cluster Communication Cost f (∅) = 0

Calculate Intra Cluster Communication Cost f (I)

Orbital to orbital distance from Algorithm 2 when   mCCA is starting

Take some nodes position Pi where mCCA starts to spiral propagation within Si

Check f(∅) and f (I) after some specific time slot

IF f (∅) = f (I) **THEN**

　　Stop spiral propagation

　　Insert into CS (cluster set)

ELSE

　　Carry on spiral propagation

END IF

Start data aggregation and data transmission

END WHILE

CCABC (Algorithm 2): Orbital to Orbital distance Calculation

Set r_{ca} is current transmission range

Set $r_{ca} = r_{max}$

Finding neighbour (r_{ca}, **i, j**)

Calculate D (Density)

Calculate r_{min}

Calculate r_{od}^2

Set the orbital distant r_{od}^2

CCABC (Algorithm 3): Verification Algorithm

IF nodes Energy is less than threshold **THEN**

　　Nodes is dead

　　IF Ti = 0 **THEN**

　　　　CCA Is rotated

　　ELSE

　　　　Decrement time

　　END IF

END IF

In proposed scheme all nodes have two states; the active state and the stand-by state. In active state a node senses the data and transmits it to cluster head. The stand-by nodes are in sleeping mode. Any one node from the nucleus is acting as cluster head. Each sub-orbit transmits their data to the upper sub-orbit within an orbit (Fig. 2). In each orbit, the sub-orbit, which is nearest to the nucleus, collects other sub-orbits data. Then the data is aggregated and transmitted to next orbital. Nearest orbit of nucleus transmits data to the cluster head in nucleus. Cluster head collects data and aggregates and transmits it to the base station. The position of the base station is fixed. Orbital nodes are changing their states randomly and repeatedly.

The whole network energy as well as the energy utilization of the nodes is divided into two parts; one is data sensing and another is data transmission. In initial condition the network is divided into some clusters of actual size depending on intra-cluster communication cost f(I) and inter-cluster communication cost $f(\emptyset)$ as discussed in *Theorem 2*. The cluster nodes are arranged like atomic structure with the help of modified CCA, cluster head (CH) is selected from nucleus. At the nucleus few nodes are in active state, one out of them will act as a CH and rest will behave as normal node. At the nucleon the f(I) and $f(\emptyset)$ is minimum at this point according to *Theorem 4*. These nodes are also changing their state from active to stand-by. When CH changes its state or its energy reaches to a threshold value, any other active node from the nucleus may be designated as a CH for next round.

The transition phase is divided into two parts; one is data collection part and another is data transmission part. In the data collection part active nodes are sensing data from their monitoring area and in transmission phase selected sub-orbital nodes and cluster heads are collecting other cluster member data, aggregating and transmitting the same. In nucleus, one of the nucleon nodes is selected as a cluster head by CCABC. A sufficient amount of area is covered by the cluster head. Inter-cluster communication costs also minimum in CCABC scheme as shown in *Theorem 2*.

In CCABC, data bits are aggregated in different orbits. In CCABC model the data are propagating form lower orbit to the upper orbit. Finally cluster head gets the aggregated data from nearest orbit. Hence cluster head's data aggregation and receiving loads are distributed in different orbital node, therefore, receiving energy loss in CCABC cluster head is less. The transmitting, receiving energy losses calculating formula and CCABC orbital to orbital distance calculation formula are defined below.

V. SIMULATION RESULT

The whole simulations are done in MATLAB. The whole simulations are divided into two parts, one is the cluster formation through spiral propagation and another is data transmission with energy calculation phase. In data transmission and energy calculation phase every node checks their present energy and compares with its threshold value. If present energy is less than the threshold, then the node is declared as a dead node. The cluster formation is based on internal knowledge of each node and massage passing is not required. The spiral propagation will be stopped when the clusters are formed where the inter-cluster and intra-cluster communication cost is minimum. The nodes are sensing data, aggregating and transmitting it to the cluster head. In each round, cluster member nodes transmit their sensed data for five times to the cluster head. The CH collects this data and then it is transmitted after aggregation to the base station. After data transmission round each node calculates their remaining energy according to transmission and receiving energy loss equations which is describe in [19]. These two processes are running continuously for the entire life span of the network.

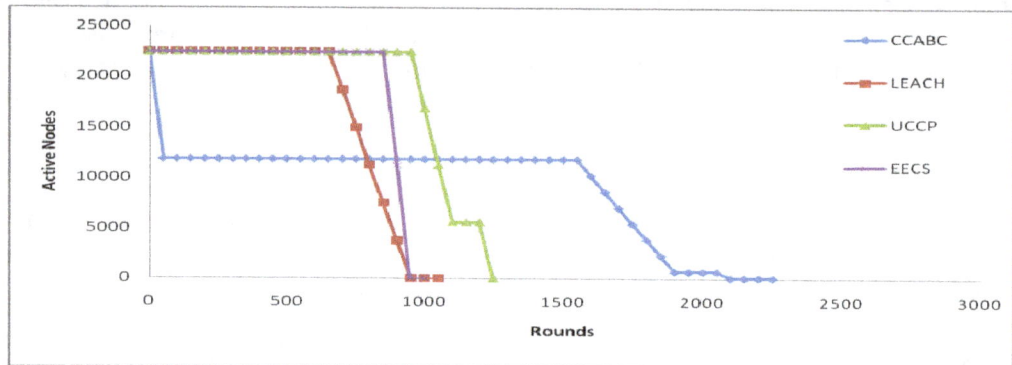

Fig. 6: Number of Active Nodes in CCABC

We have taken 150×150 matrix and number of state k=15. In this matrix we have applied mCCA. The CCABC nodes are self-organized. The Fig. 6 shows the total number of active nodes per round in a network. In the CCABC, sensor node survives longer time than the other automata based algorithm and cluster based algorithms e.g. LEACH, UCCP, M-LEACH and EECS. In this CCABC, clusters are generated without message transmitting and cluster heads are selected on the basis of local information. Whereas, the other popular clustering technique like LEACH, UCCP, EECCP some amount of energy are spent by the message passing for cluster generation. In CCABC, message over heading problem can be recovered, which is present in LEACH clustering technique. Table 1 show the parameter values which is used in simulation.

Table 1

Simulation parameters

Sensor Deployment Area	150×150
Base Station Location	(50,175)
Number of node	22500
Data Packet Size	800bit
Initial Energy	0.5J
Stand-by state energy loss	0.00006J
Energy per bit spent by the transmitter circuit (e_t)	50 nJ/bit
Amplifier energy (e_d)	10 pJ/bit/m^2

In CCABC simulation technique, initial energy of a node is 0.5J and packet size of each message is 100 bytes is fixed in whole simulation. The base station position is fixed. In each round orbital nodes collect data and send message with the help of TDMA (Time division multiple-access) MAC protocol [18]. In each round, nodes are sending five data message to the cluster head. Cluster head collect their data and send data message to the base station. For data aggregation, energy spent is 5nJ/bit/message [2].

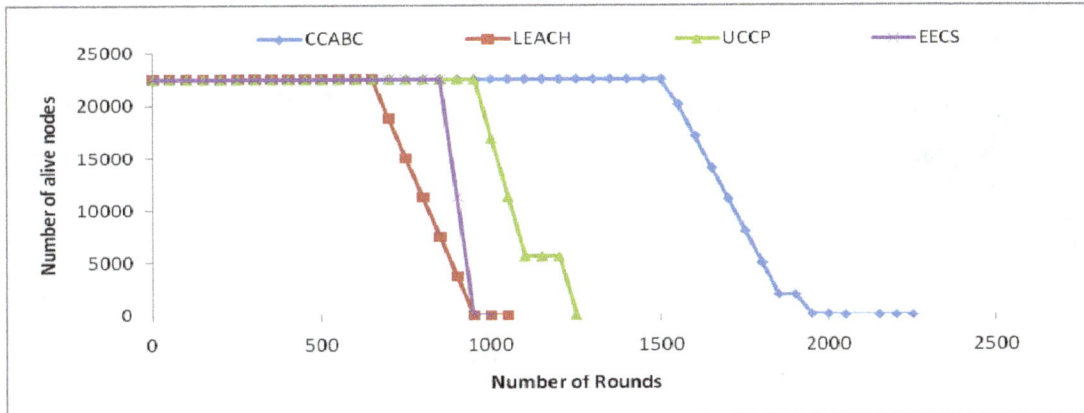

Fig.7: Life time of the nodes in CCABC

When the clusters are generated by CCABC scheme at the beginning, the active nodes are 10652 in number. These active nodes cover whole network field, and number of stand-by nodes are 11848. We compare CCABC with other clustering algorithm LEACH, UCCP (Unified Clustering and Communication Protocol [5]), EECS (Energy Efficient Clustering Scheme [6], [14]. The result shows that (Fig. 7.) in CCABC, death of the first node occurs after 1560 round which is better than other existing algorithm. The CCABC extend network life time 41% over UCCP, 64% over LEACH and 54.55% over EECS.

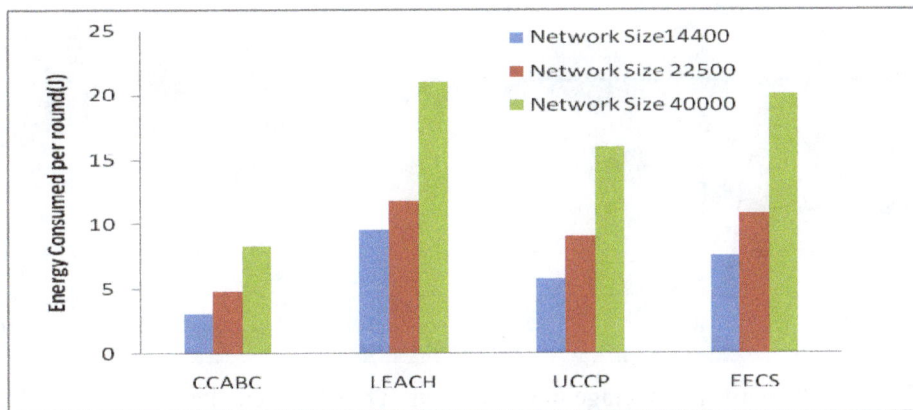

Fig.8: Average energy consumed per round

The Fig. 8 represents the average energy consumption per round for three different network sizes. These statistics are collected using 1500 independent rounds with no dead nodes in the network. It can be observed that CCABC outperforms over all other protocol because it generates clustering with the help of cyclic cellular automata and a large number of nodes are in stand-by state. On the other hand CCABC reduces message overheads compared to other technique.

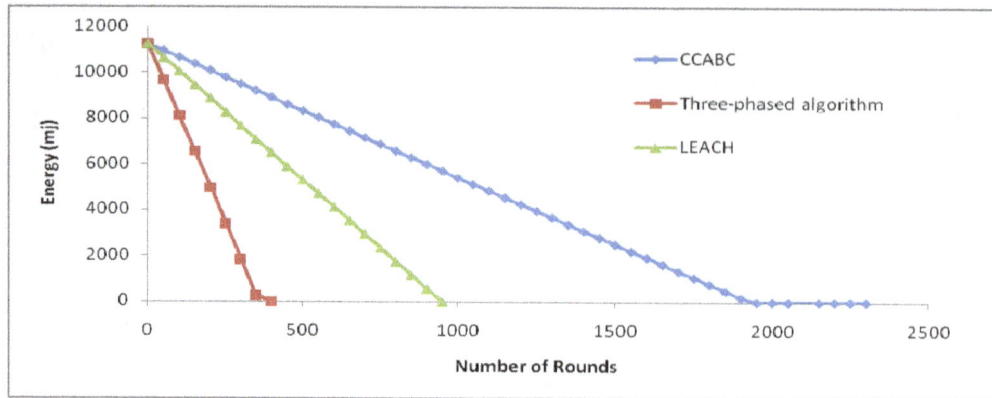

Fig. 9: Total amount of energy of the network in energy CCABC

Total energy spent by the sensor network in CCABC, three-phased algorithm and LEACH are compared in Fig. 9. The CCABC saves 60.5305(J) energy per round compared to three-phased algorithm and 13.898(J) energy per round compared to LEACH. The better energy utilization of CCABC scheme increases network's lifetime 81.83 %   more compared to three-phased algorithm and 56% more compared to LEACH.

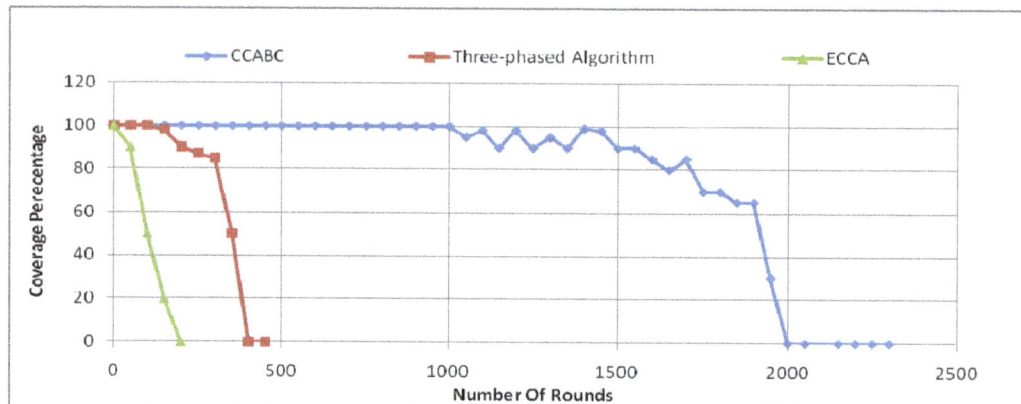

Fig. 10: The network coverage in CCABC vs. Three-phased algorithm and ECCA

The area which is monitored by the sensing range of all active nodes is called coverage area. In the Fig. 10 we have compared percentage of coverage between CCABC and three-phased algorithm. We have got better result compared to three-phased algorithm. The experimental results confirm that the network coverage is approximately 40% in three-phased algorithm and 20% in ECCA [9] algorithm, whereas, the CCABC achieves up to 80% of network coverage.

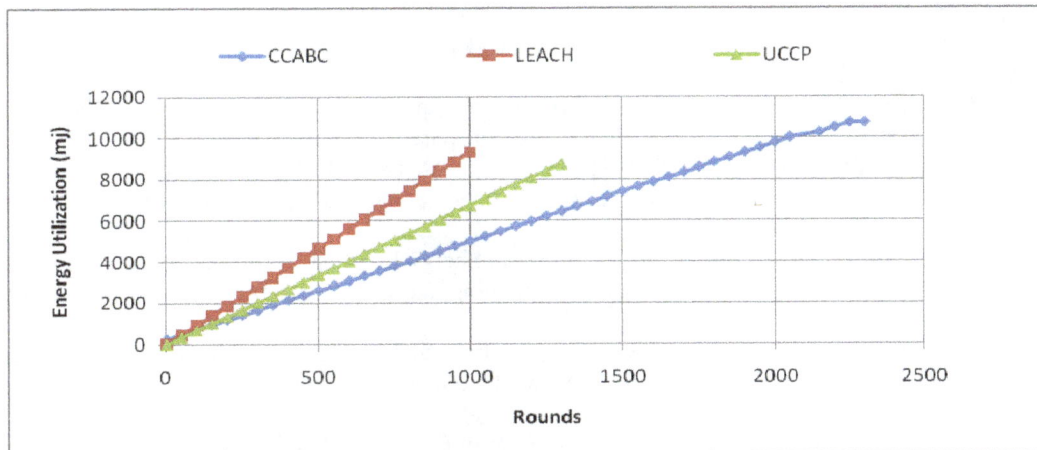

Fig. 11: Energy utilization of the sensor network

In CCABC, sensor nodes send their data with minimum energy loss compared to other existing algorithm LEACH, UCCP. In CCABC 52.82% of nodes are in stand-by state, loses minimum amount of energy. 47.34% nodes are in active state. The energy utilization for topological management in different network is shown in Fig 11. The energy utilization for topological management in LEACH and UCCP is 83% and 77.9% respectively. Whereas, in cluster based energy efficient wireless sensor networks management scheme 60.9% energy uses for topological management.

VI. CONCLUSION

In this paper we have designed an energy efficient sensor network with modified cyclic cellular automata based clustering. Modified cyclic cellular automata (mCCA) splits the whole network into some effective size clusters. The mCCA also ensure maximum coverage in the network with minimum active nodes. In CCABC we have determined an optimal cluster size where the node sends their data with minimum energy loss. An efficient cluster head position is also determined in CCABC. Here we have proposed a new effective data aggregation model which is used by the cluster heads before data propagation. The simulation results established that the proposed scheme is better compared to other popular clustering algorithms. The mobile network model works with CCABC, may show better performance in real application. The CCABC is suited for VLSI implementation and, therefore, the proposed management scheme can be implemented with low cost hardware.

REFERENCES

[1] S. Adabi, A.K.Zadeh, A.Dana, S.Adabi, "Cellular Automata Base Method for Energy Conservation Solution in Wireless Sensor Network", IEEE Wireless Communications, Networking and Mobile Computing 4th International Conference on, 2008.
[2] W.R.Heinzelman, Anantha Chandrakasan, and H. Balakrishnan, "Energy-Efficient Communication Protocol for Wireless Micro sensor Networks", IEEE 33th Hawaii International Conference on System Sciences, 2000.

[3] A. S. Zahmati, B. Abolhassani, A. A. B. Shirazi, and A. S. Bakhtiari, "An Energy-Efficient Protocol With Static Clustering for Wireless Sensor Networks", World Academy of Science, Engineering And Technology 28 2007.

[4] W. Heinzelman, A. Chandrakasan, H. Balakrishnan, An application specific protocol architecture for wireless microsensor networks, IEEE Transactions on Wireless Communications 1 (4) (2002).

[5] N. Aslam, William Phillips and William Robertson, "A Unified Clustering and Communication Protocol for Wireless Sensor Network", IAENG International Journal of Computer Science, 35:3, IJCS_35_3_01, 21 August 2008.

[6] M. Ye, C.F. Li, G.H. Chen, and J. Wu, "EECS: An Energy Efficient Clustering Scheme in Wireless Sensor Network," IEEE Int'l Performance Computing and Communications Conference (IPCCC), pp. 535-540, 2005.

[7] S. Lee, J. Lee, H.Sin, S.Yoo, S. Lee, J.Lee, Y. Lee, S. Kim, "An Energy Distributed Unequal Clustering Protocol for   Wireless Sensor Networks", World Academy of Science Engineering And Technology 48, 2008.

[8] H.Long, Y.Liu, X.Fan, R.P.Dick, H.Yang, "Energy-Efficient Spatially-Adaptive Clustering and Routing in Wireless Sensor Networks", EDAA, 978-3-981080-5-5/DATE09, 2009.

[9] R.O.Cunha, A.P.Silva, Antonio A.F.Loreiro, Linnyer B. Ruiz. ,"Simulating Large Wireless Sensor Network Using Cellular Automata", In IEEE Proceedings of the 38th Annual Simulation Symposiums (ANSS'05), pp. 313-330, April 2005.

[10] I. Banerjee, S. Das, H. Rahaman and B. K. Sikdar, "An Energy Efficient Monitoring of Ad-Hoc Sensor Network with Cellular Automata", IEEE International Conference on System Man and Cybernetics. Oct 8th – 11th 2006 Taiwan

[11] Y.M.Baryshnikov, E.G.Coffman, K.j.Kwak, "High Performance Sleep-Wake Sensor System based on Cyclic Cellular Automata", IEEE International Conference on Information Processing in sensor Networks, 2008.

[12] C. R. Shalizi, K.L. Shalizi, "Quantifying   Self-Organisation in Cyclic Cellular Automata ", Proceedings of SPIE, vol 5114, 2003.

[13] J. Greenberg and S. Hastings, "Spatial patterns for discrete models of diffusion in excitable media" SIAM Journal of   Applied Mathematics", 34515-523, 1978.

[14] R.Fisch, J. Gravner, and D. Griffeath. "Metastability in the Greenberg-Hastings model", The Annals of Applied Probability, 3(4):935-967, 1993.

[15] S. Liao, " On the general Taylor theorem and its applications in solving non-linear problems", Communications in   Nonlinear Science and Numerical Simulation, Volume 2, Issue 3, September 1997, Pages 135-14.

[16] Hui-Ching Hsieh, Jenq-Shiou Leu, Wei-Kuan Shih, "A fault-tolerant scheme for an autonomous local wireless sensor   network", Computer Standards & Interfaces, Volume 32, Issue 4, June 2010, Pages215-221.

[17] S. Pattem, B. Krishnamachari, and R. Govvinadan, "The impact of spatial correlation on routing with compression in wireless sensor network", in Proc. Int. Symp. Information Processing in sensor Network, Apr.2004, pp.28-35.

[18] Wei Ye, John Heidemann, Deborah Estrin, "Medium Access Control with Coordinated Adaptive Selling for Wireless Sensor Network", IEEE Transation on Networking, 2004.

[19] Indrajit Banerjee, Prasenjit Chanak, Biplab k. sikdar, Hafijur Rahaman, "EERIH: Energy Efficient Routing via Information Highway in Sensor Network" IEEE International conference on emerging trends in Electrical and Computer technology, March 23rd and 24th 2011, kanyakumari, India.

14

POWER EFFICIENT CLUSTERING PROTOCOL (PECP)- HETEREGENOUS WIRELESS SENSOR NETWORK

S.Taruna[1] , Kusum Jain[2], G.N. Purohit[3]

Department of Computer Science, Banasthali University, Rajasthan, India
[1]staruna71@yahoo.com
[2]Kusum_2000@rediffmail.com
[3]gn_purohitjp@yahoo.co.in

ABSTRACT

In this paper we propose a new Power Efficient Clustering Protocol(PECP) for prolonging the sensor network lifetime. Homogeneous clustering protocols assume that all the sensor nodes are equipped with the same amount of energy and as a result, they cannot take the advantage of the presence of node heterogeneity. Adapting this approach, we propose a new protocol named PECP (Power Efficient Clustering Protocol) to improve the stable region of the clustering hierarchy process using the characteristics parameters of heterogeneity, namely the fraction of powerful nodes(with more energy) more suitable to become a cluster head. Intuitively, powerful nodes have to become cluster heads more often than normal nodes. The performance of the PECP via computer simulation is evaluated and compared with other clustering algorithms. It has been found that PECP prolongs the sensor network lifetime.

KEYWORDS

Clustering Protocols, homogeneous, Network Lifetime , Cluster head, heterogeneity.

1. INTRODUCTION

A sensor network consists of a large number of very small nodes that are deployed in some geographical area. These tiny sensor nodes, which consist of sensing, data processing, and communicating components, leverage the idea of wireless sensor networks [1], [2]. As sensor nodes may be placed everywhere, this type of network can be applied to multiple scenarios [3]. e.g., in healthcare [4], where they are used to monitor and assist disabled patients, habitat monitoring [5], disaster management [6], and even for commercial applications such as managing an inventory, monitoring product quality, surveillance, and target tracking [7].

In cluster based architectures, mobile nodes are divided into virtual groups. Each cluster has adjacencies with other clusters. All the clusters have the same rules. A cluster can be made up of a Cluster Head node and Cluster Members [8]. In this kind of network, Cluster Head nodes are used to control the cluster and the size of the cluster is usually about one or two hops from the Cluster Head node. There are many cluster based architectures [9]. Sensor networks clustering schemes can be classified according to several criteria. For example, they can be classified according to whether the architectures are based on Cluster Head [10] or on Non Cluster Head [11]. The first architecture needs a Cluster Head to control and manage the group, and the second one does not have a specific node to perform this task. Another way to differentiate the cluster-based architectures is observing the hop distance between node pairs in a cluster.

The concept of cluster based routing is also utilized to perform energy efficient routing in WSNs. Author[12] shows the most important features of cluster-based architectures over ad hoc and sensor networks. The last feature is strongly linked with energy conservation, given that clustered wireless sensor networks offer two major advantages over their non-clustered counterparts; firstly, clustered wireless sensor networks are capable of reducing the volume of inter-node communication by localizing data transmission within the formed clusters and decreasing the number of transmissions to the sink node; secondly, clustered wireless sensor networks are capable of extending the nodes' sleep times by allowing cluster heads to coordinate and optimize the activities. Some of routing protocols in this group are: LEACH (13), PEGASIS (14).

Researchers generally assume that the nodes in wireless sensor networks are homogeneous, but in reality, homogeneous sensor networks hardly exist. In heterogeneous sensor networks, typically, a large number of inexpensive nodes perform sensing, while a few nodes having comparatively more energy to perform data filtering, fusion and transport. This leads to the research on heterogeneous networks where two or more types of nodes are considered. Heterogeneity in wireless sensor networks can be used to prolong the lifetime and reliability of the network.

For heterogeneous WSNs, a very critical task for clustering protocols is to select the cluster head so that least energy is consumed, and hence prolong the lifetime. Clustering algorithms can be classified based on two main criterions: according to the energy efficiency and stability. Selection of cluster head in energy efficient techniques generally depends on the initial energy, residual energy, average energy of the network, or energy consumption rate or combination of these. The stable election protocols for clustered HWSN prolong the time interval before the death of first node called stability period. Many of the algorithms are proposed for clustering in heterogeneous network. Following are the some algorithm for clustering: EEHC[15], DEEC [16], DBEC 17] and CBSD [18].

The main contribution of our work is the design and verification of a new clustering protocol for heterogeneous sensor network and its comparison with other protocols in existence. In this paper, we present a proposal for new cluster head selection. The proposed algorithm is zone based algorithm and heterogeneity parameter is energy. In this algorithm we reduce the number of communication between the sensors nodes for cluster head selection, so that the energy consumption for cluster head selection can be further reduce.

Rest of the paper is organized as follows. Section 2 formulates the problem, explains which issue has to be solved and presents some application environments where our proposal can be used. Section 3 describes our proposal. Its scalability is demonstrated in Section 4.Section 5 describe the energy model. Section 6 gives the proposed algorithm and operation. The comparison of the proposed protocol with other existing cluster based protocol is shown in Section-7. Finally, Section 8 gives the conclusion.

2. PROBLEM FORMULATION AND APPLICATION ENVIRONMENTS

Wireless sensor network generally have the architecture presented in [1]. Let us suppose an environment where a great variety of sensors must be scattered to take measurements from the environment. Let us also suppose that the area is divided into rectangular zones. The network has two types of nodes: Cluster Head (CH) and cluster member (CM). The CM senses the data and send to CH in specific time duration or when an event happens in the network. CH is responsible to send data to Base Station (BS). BS is situated far and out of network. CH is also responsible for data aggregation and data compression that reduces the energy consumed for data transmission from CH to BS.

Following type of communication will take place in the network:-

1. CM senses the data from the environment and sends it to CH.
2. CH aggregate the data come from CM and compress the data and send it to BS.

3. Control information from BS to CM and CH.
4. Control information from CM and CH to BS.

Taking into account the aforementioned premises, several application environments can be found. Some of them are the following:

1. Applications with the fixed deployment area. In which the area can be divided in rectangular grid. For example like WSN for fire detection. It gives the good results work with applications in which node are fixed. For example WSN for fire detection.
2. Application areas where the movement of node is less or symmetric. In this the means of symmetric is: the density of zone should not be so much affect by the movement of node.
3. It could be used in any kind of system where an event or alarm is based on what is happening in a particular zone of network. One example is a group-based system to measure the environmental impact of a place (forest, marine reef, etc.).
4. Agricultural application for example some harvests like saffron: a small change in the temperature can be damage the plants or flower of the saffron.
5. WSN for Traffic Management.

3. ARCHITECTURE PROPOSAL DESCRIPTION

From the logical point of view, let us suppose an environment where a large number of same type of sensors are scattered to take measurements from the environment. The area is divided into rectangular grid called zones and each zone has several sensors sensing the environmental value like humidity, temperature, wind, movement, etc., organized by a central node as a cluster head (CH). The network has 2 type of node cluster heads (CHs) and Cluster members (CMs). Each cluster head have an association with some sensing node called Cluster Members (CMs). Although CHs have sensing capacities, they are the ones with higher capabilities then CMs. CMs only sense the environment and send the sense data to the CH. The same physical and MAC layers are used between CHs and their CMs. CHs organize and control the CMs in their cluster and all CMs have to establish a connection with a CH to join its cluster. This connection can be established only if the distance between them is shorter than other CHs in the network. All CHs and CMs can communicate with the BS. In our design only one CH per cluster has been provided, but more could be added for scalability purposes, using the algorithm presented by the same authors of this paper in reference [25]. An example of architecture proposed is shown in figure 1. The area is geographically divided in zone.

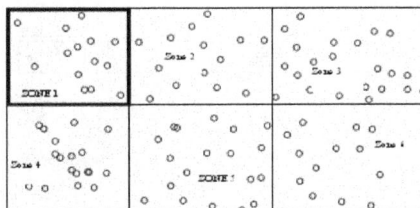

Figure 1: wireless sensor network

Many of the routing protocol can be used to route information between CHs. For Example OLSR [19] , AODV [20], DSR [21] or TORA [22].   The number of clusters in the network is determined by the extension that is to be covered by the whole network. If a new zone needs to be covered, a new cluster has to be added. Although many types of sensors or types of devices can be added to any cluster, the application of the 20/80 rule (20% of CHs, 80% of CMs) is suggested [23].

3.1. Identifiers and Predefined Parameters

Base Station (BS) has the list of zone with an identifier called ZoneID and geographical coordinate of zone in network area. Each node deployed in the network with a unique 16-bit identifier called nodeID. When a node in a zone declared as cluster head by Base Station (BS) it acquires a unique 16-bit cluster head identifier called clusterID with zoneID in which it will work. Base station also declared a threshold value (TH). The selection of the new cluster head or for the CH election algorithm runs in the system when any of CH has its residual energy less or equal to threshold value. All nodes in a cluster will have the same cluster head and any new node will join the cluster whose CH is closest.

Every node in networks must have the following parameters in the proposed architecture:

- **Position:** position is the (x,y) geographical value for the node. It could be given manually or by GPS.
- **Type:** It identifies the type of node, whether it is CH or CM.

With above parameter when a node is declared as CH, following parameters are added by Base station to node.

- **Max_con:** Maximum number of supported connections from CMs.
- **Max_distance**: It is the maximum distance to be a neighbour. It is always shorter than or equal to the coverage area radius. The energy model as stated above specifies that distance can be calculated by signal strength. Maximum distance is the distance for transmission of request or information from the CH.

Two parameters have been defined to be used for the operation of the architecture.

Δ **Parameter:** It depends on the node available energy and its age in the system. It is used to ascertain which node is the best one to become a Cluster Head node. A node with higher residual energy and the most stable node will be the CHs. Since the oldest node should be the lowest energy node, but this parameter appears to consolidate the most stable nodes as the CHs (new ones could be mobile nodes or even with lower energy). A node with higher available energy and older will have higher Δ. Equation 1 defines the Δ parameter:

$$\Delta = 16\text{-age } (1\text{-}E^2)^{1/2} \tag{1}$$

where *age* = log2(*nodeID*), so age varies from 0 to 16. *E* is defined as % of energy consumption and it values vary from 0% to 100%. *E* = 0 indicates it is fully charged and *E* = 100 indicates it is fully discharged.

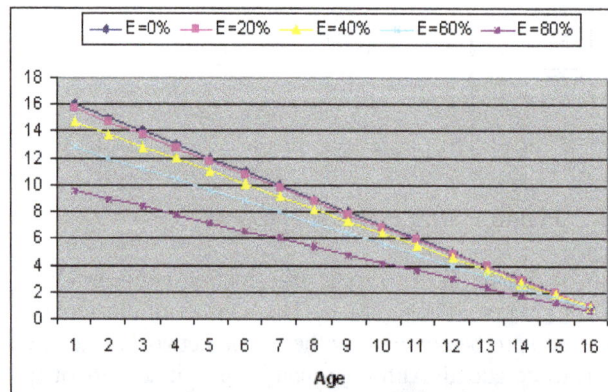

Figure 2: Δ values versus age

ρ Parameter

It is the property of a node to become a cluster head. It is used by the Base Station or election algorithm to elect a node as CH. When energy of any CH in the network is less then threshold value CH election algorithm will run in the system by the base station or by the existing CH. CH election algorithm is run in the following scenario:-
1. For the first time when the network is deployed the election of CH is done by the BS.
2. When any of the existing cluster head is dead or non working with any reason then election algorithm is done by the base station.
3. When the Δ parameter of existing CH is equal to TH then election algorithm is run by the existing CH to select a new cluster head.

The ρ parameter depends on the CH with the residual energy. If the residual energy of the CH < = Threshold, value of ρ will be 0. Value of ρ will be 0 or 10, according to the state of the node. A 0 value of ρ indicates that the node is not elected as CH till now and a 10 value of ρ means it has the energy level at threshold value and not be elected till all node has the same ρ value. If all the nodes has same value of ρ (ρ=10), than a node will be selected as cluster head with higher Δ.

4. SCALABILITY

It is known that cluster based systems are more scalable than other systems. This section shows that why our proposal scales better than other proposals. First, we have to take into account that computation is much cheaper than communication in terms of energy dissipation [24]. So, what is desired is architecture with fewer retransmissions. This will imply a saving in energy of the whole system and it will give more scalability to the architecture.

Let a network of nodes G = (V, E) be, where V is the set of nodes and E is the set of connections between nodes. Let k be a finite number of disjoint clusters of V, so V = U (V_k) and there is no node in two or more subsets ($\cap V_k = 0$), i.e., there are not overlapping nodes. Let us suppose N = |V| (the number of nodes of V) uniformly distributed in a region. Let us suppose that there is just one cluster head node per cluster, so there are k head clusters in the whole network. Equation 2 gives the number of nodes:

$$N = \sum_{i=1}^{k} |V_k|$$

(2)

and the average number of members of in a cluster will be given by Equation 3:

$$Average = \frac{N}{k}$$

(3)

By the architectural proposal and the above formula the average number of members in cluster is as in equation 3. The minimum number of neighbouring zone for a zone in network is 3 (If a zone is at corner of geographical area) and maximum of 8 neighbouring zone in the network (if it is in middle of the geographical area).

5. ENERGY MODEL

We assume that the energy consumption of the sensor is due to data transmission and reception. Cluster head is also consuming energy for the data aggregation before it sends the data to BS. We use the same radio model as stated in [13], [24] and shown in Figure 1, the energy consumed in transmitting one message of size *k* bits over a transmission distance *d*, is given by

$E_{Tx}(k,d) = k(E_{elec} + \Delta_{AMP}d^{\square}) = E_{elec}\,k + k\Delta_{AMP}d^{\square}$

Where k=length of the message

d=transmission distance between transmitter and receiver

E_{elec}= electronic energy

Δ_{AMP} =transmitter amplifier

Δ = Path Loss (2<= Δ <=4)

Also, the energy consumed in the message reception is given by

$$E_{rx} = E_{elec}\, k$$

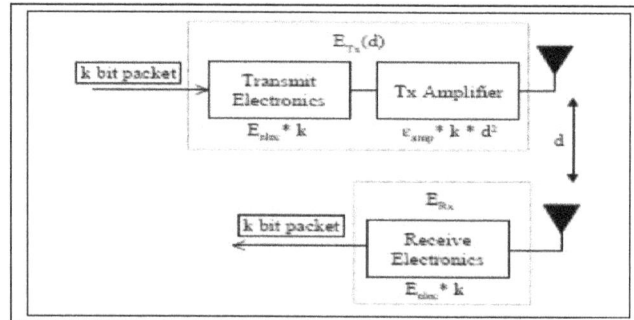

Figure3 : Radio Energy Model

6. PROPOSED ALGORITHM AND ARCHITECTURE PROPOSAL OPERATION

We proposed a new cluster head selection protocol for heterogeneous wireless sensor network. The algorithm is based on LEACH protocol. The proposed algorithm is zone-based algorithm and the parameter for heterogeneity is energy. In wireless sensor network two basic approaches are used to select the cluster head: Centralized and Distributed. In either of the method after a node is declared as a cluster head , the following process is followed to associate the node to cluster head:

1. The node declared as cluster head send the JOIN request to all other nodes.
2. The node receives the JOIN request and examines it on the basis of signal strength. According to general radio energy model for wireless sensor networks the signal strength depends on the distance means if signal comes from the more distance has the less strength and vice-versa.
3. A node will be selected as the cluster head, which has minimum distance from node.
4. Node sends a message for joining the cluster as cluster member.

The probability of the node to become a cluster head will be P, p is the maximum number of cluster heads in the network. In the algorithm we reduce the number of communication between the sensor nodes for cluster head selection so that the energy consumption for cluster head selection can be further reduce. This can increase the residual energy of the cluster head and the network survivability can be enhanced. The emphasis of our approach is to increase the life span of the network by ensuring   a uniform distribution of nodes in initial network. In the following section we describe the propose algorithm. Let us suppose a heterogeneous network with 100 nodes having different initial energy. The base station (BS) is located outside the deployed area and fixed. According to energy model data compression energy is different from the reception and transmission. Energy of transmission depends on the distance (source to destination) and data size. Although nodes are mobile but during the cluster head selection phase the nodes are assumed to be immobile.

The proposed algorithm works in round. Each round has the setup and transmission phase.

The flow graph for the proposed algorithm is as follows :

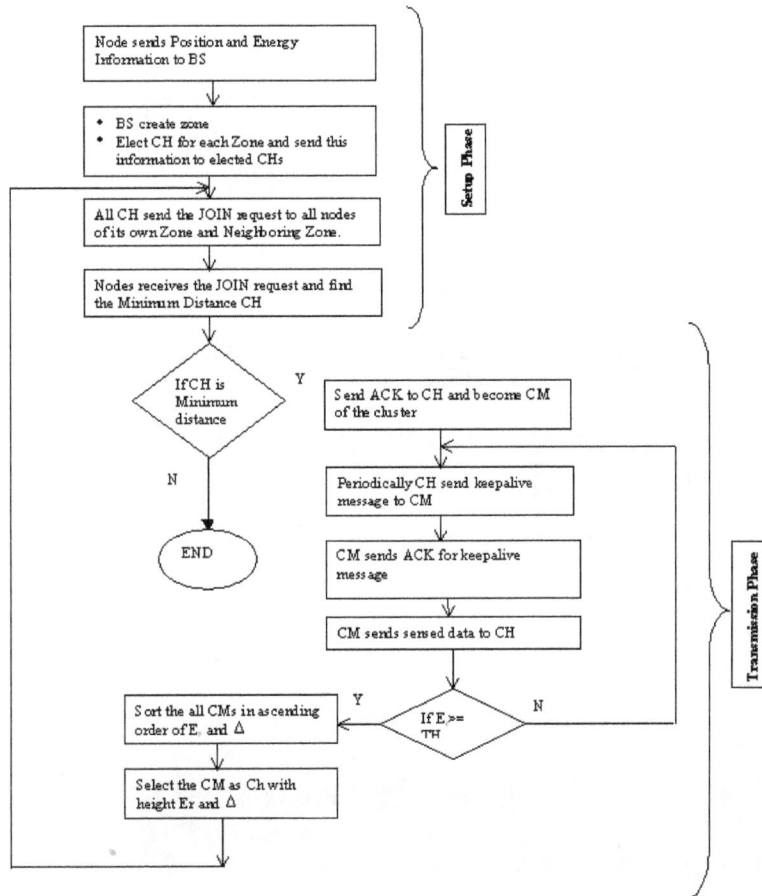

6.1 Setup phase

In the setup phase cluster head selection is done. Following steps are will be followed.

6.1.1 Creating Zone

- All nodes send the energy and position information to BS in SEND_INFO message.
- BS create zone on the basis of geographical area and assign a zoneID. For example we divide the network in 6 rectangles.

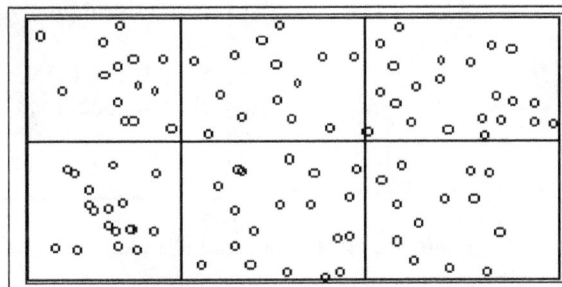

Figure 4 : The sensor network

- By the position value it select node in zone and run cluster selection algorithm to select the cluster head.

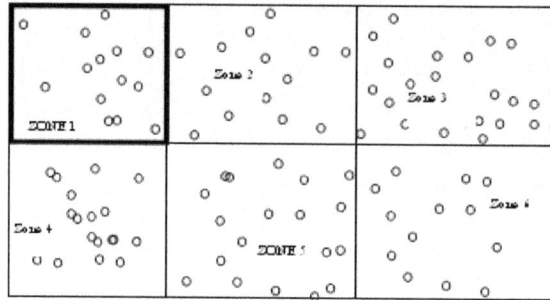

Figure 5: Zone in the network.

- BS select CH elect Message node, elected as cluster head to cluster with clusterheadID and zoneID with which it associated by ELECT_INFO message.

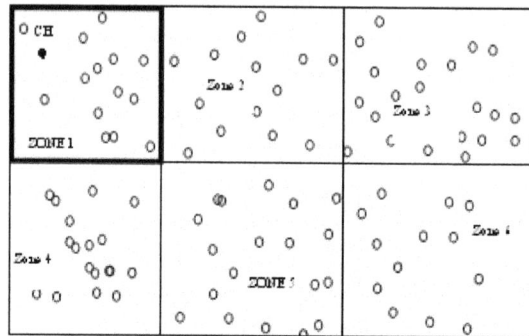

Figure6 : Selection of Cluster Head in Zone 1

Protocol operation to create zone by BS:

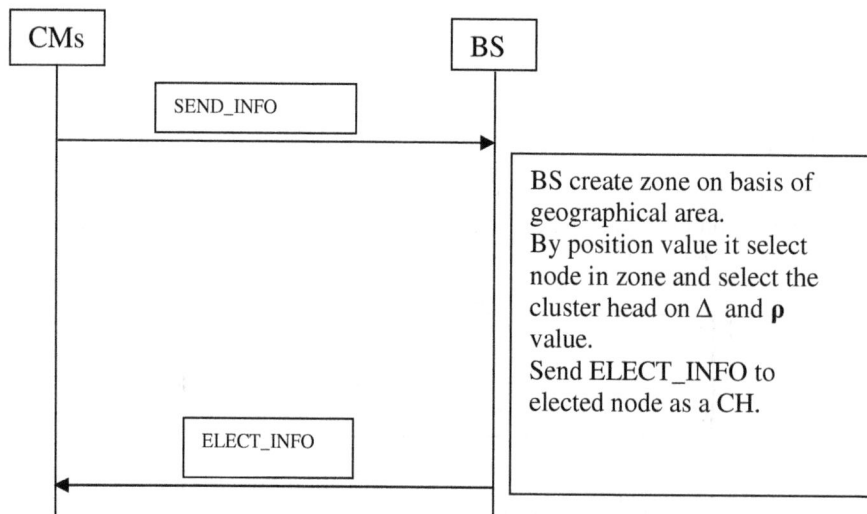

Figure 7 : Protocol operation to create zone and CH election by base station

In this process two messages are used :-
1. SEND_INFO (Energy and position information of Node)
2. ELECT_INFO.(ClusterID with ZoneID from which CH is associated).

6.1.2. Cluster Formation.

- Selected cluster head send the JOIN request to the nodes lie in the zone itself and to the neighbour's zone. For example in figure the cluster head of Zone 1 will send the join request to the nodes of Zone 1, Zone 2, Zone4, and Zone 5.

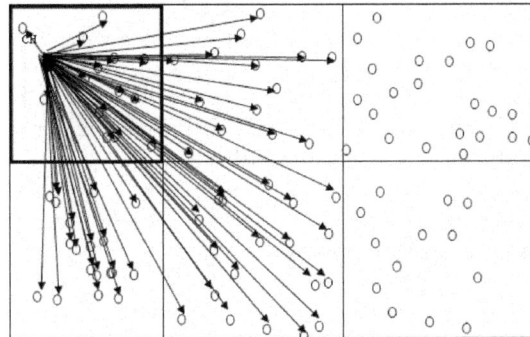

Figure 8 :Join request send to all other nodes in neighbouring zone

- Node receives all JOIN request and by their energy calculates the distance of the cluster heads.
- The node sends an ACK in response to the JOIN to the cluster head which has minimum distance.

Protocol for cluster formation

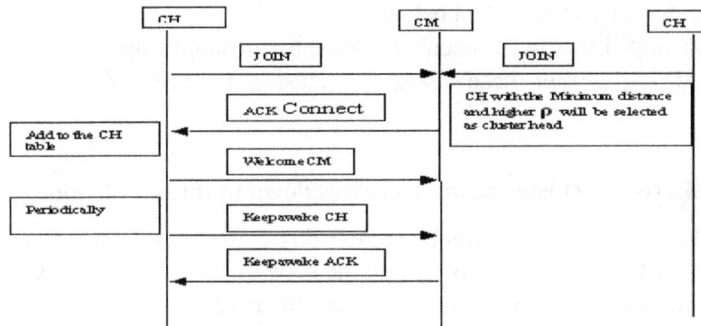

Figure 9 : Join request send to all other nodes in neighbouring Zone

- Final cluster formation is done.
- After welcome CH the cluster head periodically send the keepawake CH message to the node to check the whether the node is alive or not.
- The node responds this by keep awake ACK message to inform CH that it is in working condition.

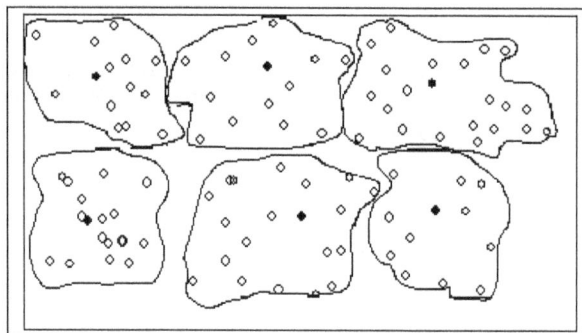

Figure 10 : Final cluster formation

The whole operation use the following Messages:
3. JOIN (cluster ID)
4. ACK connect(ClusterID, nodeID)
5. WelcomeCM(Cluster ID)
6. Keepawake(Beacon)
7. KeepawakeACK(Beacon)

6.2 TRANSMISSION PHASE

1. Node sends data to cluster head in each round.
2. Cluster head receive the data, aggregate it and send it to the Base Station.

6.3 NEW CLUSTER FORMATION

New cluster is selected for the reasons described below
- When the energy of existing cluster heads decreases down to threshold value.
- When any of the existing cluster head leave the cluster, in case of failure or disconnection.

6.3.1. When the energy of existing cluster heads decreases down to threshold value.

In this situation the election algorithm is run by the existing cluster head. And after the election algorithm the cluster head will be the part of the network as normal node (CM).
- Nodes of the existing cluster are sorted in ascending order of energy.
- Node with the highest energy and with R value
- The hop distance from the existing cluster head is minimum (hop distance may 1 or 2).
- The new selected CH will again perform the step of setup phase to select to form a new cluster.
- Old CH send all the control information by Backup Message and cluster ID.
- New selected CH send the cluster ID to the BS with Zone ID with it nodeID in N_INFO message
- The new selected node send again send the JOIN request to all node in its own zone and to neighbour zone node.
- The node will select the clusterhead with minimum distance.
- With the ConnectACK a node can select the Cluster .
- Welcome CH is send by cluster head.
- And periodically it send the keepawake message to node which is responded by the keepawake ACK by node.

- New Cluster with new CH will be created.

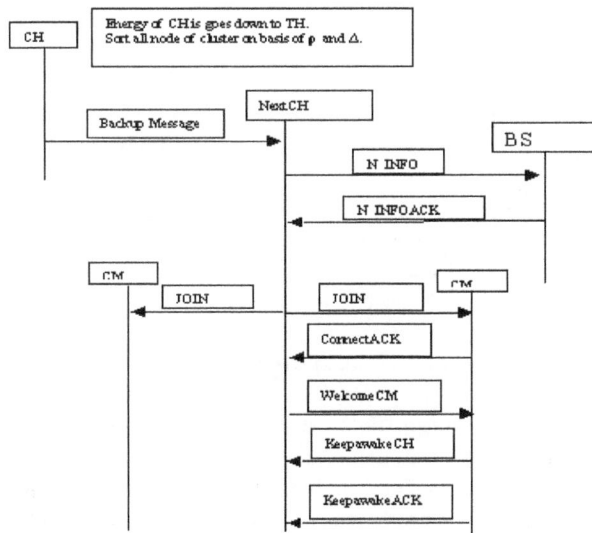

Figure 11 : Protocol operation for selecting new cluster head by existing cluster head

The whole process use the following new messages:-
8. Backup Message (nodeID,Cluser ID Associated with the CH)
9. N_INFO (Cluster ID , nodeID, zone ID)
10. N_INFOACK(ACK of n_INFO).

6.3.2 New cluster head selection in case of Failure or disconnection.

In this case the cluster head is not working in network due to failure or disconnection of CH. This time the election algorithm is run by BS and it is same as describe in 6.1.

6.4 ADDITION OF NEW NODE IN NETWORK

- When a new node is added to the network it sends it geographical information to the BS.
- Then BS send it to the zoneID to the node.
- In order to join the architecture, the new node broadcasts a discovery message.
- CHs will reply with a discovery ACK message with their position .

Node will select the CH with following three ways.

6.4.1 If it does not receive any reply within 10 seconds, it becomes a CH, so it creates the cluster and waits for new nodes. Ten seconds have been chosen because it is enough time to receive a reply from a near node. Later replies will be from nodes which are either too far or too busy. Then it send Discovery message to BS. BS station replies with ACK and associated zoneID and cluster ID of neighbouring zone. Then it send discovery message again to those node. If it receives one or more reply it follow the 6.4.3.

6.4.2 If it receives one or several replies and it finds the CH with minimum distance. Then, it sends an M connect message to establish a connection with the selected CH. If the CH confirms that connection because it has not reached the maximum number of connections, it adds this entry to its CM table, sends a —Welcome CH message.

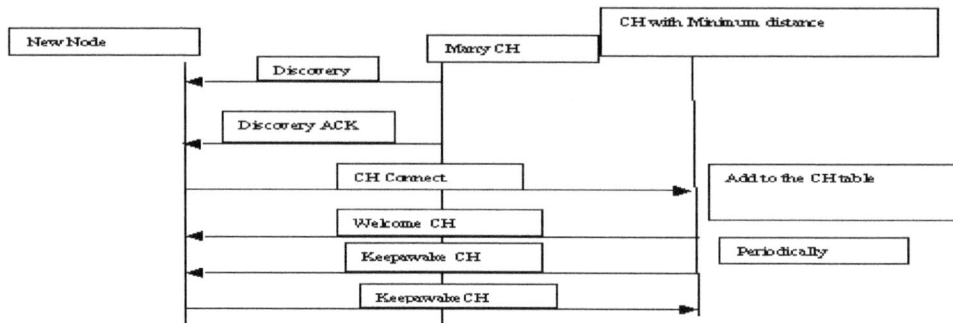

Figure 12: Protocol operation to select new cluster head by node

6.4.3   If it receives one or several replies and it finds the CH with minimum distance. Then, it sends an M connect message to establish a connection with the selected CH. If the CH does not confirms that connection because it has reached the maximum number of connections, newly added node will send the M Connect message to second best Ch and follow the same steps.

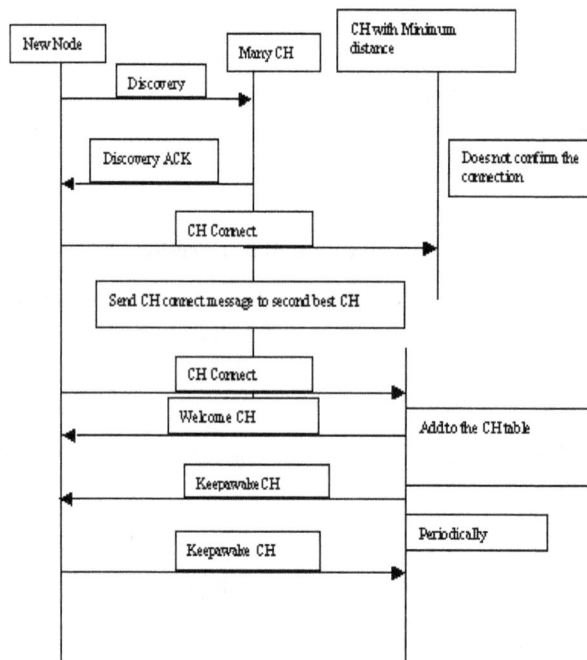

Figure 13 : Protocol operation to select new cluster head

The process uses the following message:
    11. Discovery (nodeID)
    12. DiscoveryACK(ClusterID, zone ID,nodeID).

6.5 Deletion of node in network

When a CM leaves the cluster (because of its mobility, or due to failure or other issues), the CH of its cluster will not receive any —keepawake message from it. After a dead time, the CH will erase this entry from its CM table.

6.6 Simulation

Simulation result

The performance analysis of proposed clustering algorithm is evaluated in the MATLAB and compared with the LEACH algorithm (in which cluster selection is random selection process) in terms of the network lifetime. We simulate the proposed algorithm with the LEACH algorithm. In the LEACH random selection method in each round cluster head selection is on the basis of 1/p. not on the residual energy.

The cluster head is determined by the following function (1)

$$T(n) = \frac{P_t}{1-P_t.(r.\bmod i/P_t)^2} \begin{cases} P_t & \text{if } n \in G \\ \\ 0, & \text{Otherwise} \end{cases}$$

Where Pt is the desired percentage of cluster heads, r is the current round number; G is the set of nodes that have not been cluster-heads in the last 1/Pt rounds.
When a node is declared as cluster head it sends the JOIN request to all other nodes. In this simulation the node send the JOIN request to other 99 nodes but in the proposed algorithm the node has the maximum energy will be declared as cluster head and it send the JOIN request to the nodes of its own zone and neighbouring zone for example in the simulation if Zone 1, Zone 2, Zone 4, Zone5 have 72 nodes, then JOIN request will send to only 72 Nodes. Following energy can be saved 25 Transmission of JOIN request.

Power consumption in according to energy model method for cluster head selection and node association with the cluster head

$E_{consume}$ = JOIN request to all other nodes in network
+ Energy for the reception of data for all nodes in cluster
+ Send the CH information to Base station.
=(ETX * ctrPacketLength + Emp* ctrPacketLength)   +(ERX*ctrPacketLength *N) + EDA   +(ETX * ctrPacketLength + Emp* ctrPacketLength)

Length of ctrPacketLength is 100.

Table 1: Simulation of proposed and random algorithm

	Max no. of node in a cluster in Proposed algorithm	Maximum number of node In a cluster in random algorithm	Number of JOIN request in Proposed algorithm	Number of JOIN request in Random algorithm	Total Energy in cluster information in random algorithm	Total energy in cluster formation in Proposed algorithm
Simulation 1	25	50	76	99	75.0005	51.0005
Simulation 2	27	46	68	99	73.0005	48.0005
Simulation 3	24	38	65	99	69.0005	45.0005
Simulation 4	23	29	56	99	64.5005	40.0005
Simulation 5	21	37	69	99	68.5005	45.5005
Simulation 6	22	30	59	99	65.0005	41.0005
Simulation 7	27	32	63	99	66.0005	45.5005
Simulation 8	24	39	66	99	69.5005	45.5005
Simulation 9	26	35	72	99	67.5005	49.5005
Simulation 10	24	42	68	99	71.0005	46.5005

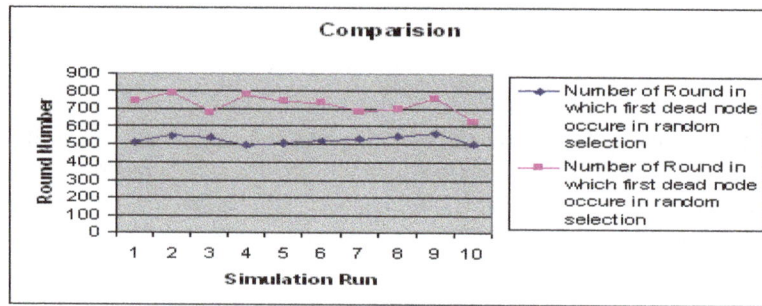

Figure 14 : Comparision operation to select new cluster head

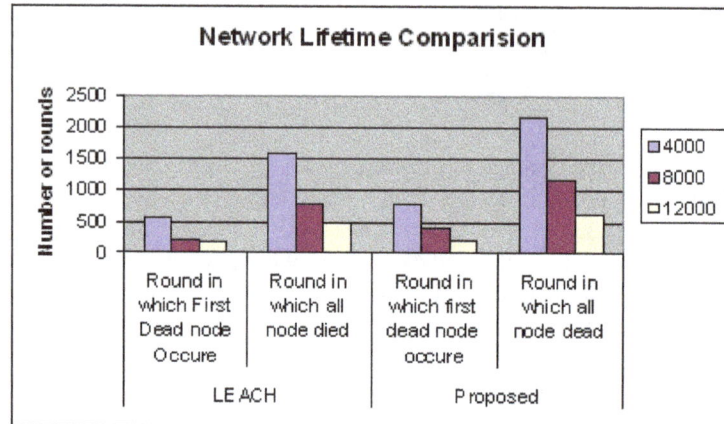

Figure 15 : Network Stability

7 PROTOCOL COMPARISION

Table 2 : Comparison of various protocols

Clustering Algorithm	Energy Efficient	Location awareness	Balanced clustering	Cluster stability	Heterogeneity Type	Clustering Methodology	Heterogeneity Level
Kumar 2009	High	No	Yes	Moderate	Energy	Distributed	Three
Qing2006	High	No	Yes	Moderate	Energy	Distributed	Two
Elbhiri 2009	High	No	Yes	Good	Energy	Distributed	Two
Duan 2007	High	No	Yes	Good	Energy	Distributed	Two
Marin-Perianu 2008	High	Yes	No	Moderate	Energy and Link	Centralized	Multilevel
Smaragdakis 2004	Low	No		Good	Energy	Distributed	Two
Haibo	Low	Yes	Yes	Very Good	Computational and Energy	Centralized	Three
Varma 2008	Low	Yes	OK	Good	Computational and Energy	Centralized	Two
Li 2007	Low	Yes	Yes	Good	Energy	Distributed	Two
Wang 2007	Ok	No	Yes	Very Good	Energy	Centralized	Two
Liaw 2009	Low	No	Yes	Good			Multilevel.
PECP	High	**Yes**	Yes	Good	Energy	Semi	Two

8 CONCLUSION

The wireless sensor networks have been envisioned to help in numerous monitoring applications. Energy efficient routing is paramount to extend the stability and lifetime of the system. In this paper, we have proposed an energy efficient heterogeneous clustered scheme for wireless sensor networks. The energy efficiency and ease of deployment make PECP a desirable and robust protocol for wireless sensor networks. In order to improve the lifetime and performance of the network system, this paper reports on the development of an architecture that creates clusters based on zone and establishes connections between sensor nodes and also the description of the protocol developed and the flow of the messages have been presented. Comparing our proposal with others, it can be seen that the detailed description of our protocol allows its easy implementation Simulations results show that PECP has extended the lifetime of the network as compared with LEACH in the presence of same setting of powerful nodes in a network. Hence, the performance of the proposed system is better in terms of reliability and lifetime. Although we compared PECP with LEACH, there are many clustering algorithms that we have to compare and there are many factors that can affect the network lifetime. Further directions of this study will be deal with clustered sensor networks with more than two levels of hierarchy.

REFERENCES

1 Charles E. Perkins. *"Ad Hoc Networking"* Addison-Wesley, Boston, MA, first edition, 2001.

2. N. Al-Karaki and A. E. Kamal. "Routing techniques in wireless sensor networks: a survey". In IEEE Wireless Communications, Volume 11, pp. 6- 28, 2004.

3. Yick, J.; Mukherjee, B.; Dipak, D. Wireless Sensor Network Survey. Comput. Netw. 2008, 52, 2292–2330.

4. Krco, S. Health Care Sensor Networks—Architecture and Protocols. Ad Hoc. Sensor Wireless Networks 2005, 1, 1–25.

5. Mainwaring, A.; Szewczyk, R.; Anderson, J.; Polastre, J. Habitat Monitoring on Great Duck Island.In Proceedings of ACM SenSys'04, Baltimore, MD, USA, November 2004.

6. Summers, S.A. Wireless Sensor Networks for Firefighting and Fire Investigation; CS526 Project. UCCS: Colorado Springs, CO, USA, 2006.

7. Yang, H.; Sikdar, B. A Protocol for Tracking Mobile Targets Using Sensor Networks. In Proceedings of SNPA'03, Anchorage, AK, USA, May 2003; pp. 71–81.

8. Jiang, M.; Li, J.; Tay, Y.C. Cluster Based Routing Protocol (CBRP). August 1998. Available online: http://tools.ietf.org/html/draft-ietf-manet-cbrp-spec-01.txt (accessed on 3 December 2009).

9. Yu, J.Y.; Chong, P.H.J. A Survey of Clustering Schemes for Mobile Ad Hoc Networks. IEEE Commun. Surv. Tutorials 2005, 7, 32– 48.

10. Lin, C.R.; Gerla, M. Adaptive Clustering for Mobile Wireless Networks. IEEE J. Sel. Areas Commun. 1997, 15, 1265–1275.

11. Ryu, J.H.; Song, S.; Cho, D.H. New Clustering Schemes for Energy Conservation in Two- Tiered Mobile Ad Hoc Networks. In Proceedings of IEEE ICC'01, Helsinki, Finland, June 2001; pp. 862– 866.

12. Abbasi, A.A.; Younis, M. A Survey on Clustering Algorithms for Wireless Sensor Networks.Comput. Netw. 2007, 30, 2826–2841.

13. W. R. Heinzelman, A. Chandrakasan, and H. Balakrishnan. "Energy efficient communication

protocol for wireless microsensor networks". In Proceedings of the Hawaii International Conference on System Sciences,2000.

14. S. Lindsey, C.S. Raghavendra, "PEGASIS: power efficient gathering in sensor information systems", in: Proceedings of the IEEE Aerospace Conference, Big Sky, Montana, March 2002.

15. Dilip Kumar a,*, Trilok C. Aseri b,1, R.B. Patel "EEHC: Energy efficient heterogeneous clustered scheme for wireless sensor networks in "Computer Communications" 32 (2009) 662–667.

16. Vivek Katiyar, Narottam Chand, Surender Soni ," Clustering Algorithms for Heterogeneous Wireless Sensor Network: A Survey ",International Journal Of Applied Engineering Research, Dindigul Volume 1, No 2, 2010

17. Changmin Duan;   Hong Fan,"A Distributed Energy Balance Clustering Protocol for Heterogeneous Wireless Sensor Networks", Wireless Communications, Networking and Mobile Computing, 2007.

18. R.S. Marin-Perianu□ , J. Scholten, P.J.M. Havinga And P.H. Hartel Cluster -based service discovery for heterogeneous wireless sensor networks International Journal of Parallel, Emergent and Distributed Systems, Vol. , No. , , 1–35.

19. Clausen, T.; Jacquet, P. Optimized Link State Routing Protocol (OLSR). RFC 3626. October 2003. Available online: http://www.ietf.org/rfc/rfc3626.txt (accessed on 3 December 2009).

20. Perkins, C.; Belding-Royer, E.; Das, S. Ad Hoc On-Demand Distance Vector (AODV) Routing. RFC 3561. July 2003. Available online: http://www.ietf.org/rfc/rfc3561.txt (accessed on 3 December 2009).

21. Johnson, D.; Hu, Y.; Maltz, D. The Dynamic Source Routing Protocol (DSR) for Mobile Ad hoc Networks for IPv4. RFC 4728. February, 2007. Available online: http://www.ietf.org/ rfc/rfc4728.txt (accessed on 3 December 2009).

22. Park, V.; Corson, S. Temporally-Ordered Routing Algorithm (TORA) Version 1, Functional Specification, Internet Draft. June 2001. Available online: http://tools.ietf.org/id/draft-ietf-manet-tora-spec-04.txt (accessed on 3 December 2009).

23 Chang, Y,C.; Lin, Z.S.; Chen, J.L. Cluster Based Self-Organization Management Protocols for Wireless Sensor Networks. IEEE Trans. Consum. Electron. 2006, 52, 75–80.

24. Heinzelman, W.R.; Chandrakasan, A.; Balakrishnan, H. Energy-efficient communication Protocol for Wireless Microsensor Networks. In Proceedings of the 33rd Annual Hawaii International Conference on System Sciences, Maui, HI, USA, January 2000; Volume 2.

25. Smaragdakis, G., Matta, I. and Bestavros, A. (2004). SEP: A stable election protocol for   clustered heterogeneous wireless sensor networks, In: Proc. of the International Workshop on

SANPA 2004. pp 251261

26. Haibo, Z., Yuanming, W., Yanqi, H. and Guangzhong, X. (In Press) A novel stable selection and reliable transmission protocol for clustered heterogeneous wireless sensor networks, Computer Communications, In Press, Corrected Proof.

27. Varma, S., Nigam, N. and Tiwary, U.S. (2008). Base station initiated dynamic routing protocol for Heterogeneous Wireless Sensor Network using clustering, Wireless communication and Sensor Networks, WCSN 2008, pp 16.

28. Li, X., Huang, D. and Sun, Z. (2007). A Routing Protocol for Balancing Energy Consumption in Heterogeneous Wireless Sensor Networks, MSN 2007, LNCS 4864: pp 79–88..

29. Li, X., Huang, D. and Yang, J. (2007). Energy Efficient Routing Protocol Based on Residual Energy and Energy Consumption Rate for Heterogeneous Wireless Sensor Networks, In: The 26th Chinese Control Conference, 5: pp 587–590.

30. Wang, X. and Zhang, G. (2007). DECP: A Distributed Election Clustering Protocol for Heterogeneous Wireless Sensor Networks, LNCS; 4489: Proceedings of the 7th international conference on Computational Science, pp 105108.

31. Liaw, J.J., D., ChenYi and YiJie, W. (2009). The Steady Clustering Scheme for Heterogeneous Wireless Sensor Networks, Ubiquitous, Autonomic and Trusted Computing, UICATC '09. Symposia and Workshops on, pp 336341.

Permissions

List of Contributors

Reena Dadhich
Department of MCA, Engineering College Ajmer, India

Aditya Shastri
Department of Computer Science, Banasthali University, India

Nivedita N. Joshi
Department of Electronics and Tele-communication Engineering, College of Engineering, Pune, India

Radhika D. Joshi
Department of Electronics and Tele-communication Engineering, College of Engineering, Pune, India

G.Sundari
Research scholar, Sathyabama University, Chennai, India

P.E.Sankaranarayanan
Dean (Academic studies), Sathyabama University, Chennai, India

Stylianos P. Savaidis
Department of Electronics Technological Educational Institute (TEI) of Piraeus, 250 Thivon & P.Ralli, Aigaleo, Athens–12244, Greece

Nikolaos I. Miridakis
Department of Computer Engineering, Technological Educational Institute (TEI) of Piraeus, 250 Thivon & P.Ralli, Aigaleo, Athens–12244, Greece
Department of Informatics, University of Piraeus, 80 Karaoli & Dimitriou, 185 34 Piraeus, Greece

Shumon Alam
Center of Excellence for Communication Systems Technology Research Department of Electrical and Computer Engineering, Prairie View A & M University, TX 77446 United States of America

O. Olabiyi
Center of Excellence for Communication Systems Technology Research Department of Electrical and Computer Engineering, Prairie View A & M University, TX 77446 United States of America

O. Odejide
Center of Excellence for Communication Systems Technology Research Department of Electrical and Computer Engineering, Prairie View A & M University, TX 77446 United States of America

A. Annamalai
Center of Excellence for Communication Systems Technology Research Department of Electrical and Computer Engineering, Prairie View A & M University, TX 77446 United States of America

Gauri Joshi
Dhirubhai Ambani Institute of Information and Communication Technology Gandhinagar, India

Prabhat Ranjan
Dhirubhai Ambani Institute of Information and Communication Technology Gandhinagar, India

D Saha
Department of Electrical Engineering and Computer Science, North South University, Dhaka, Bangladesh

M R Yousuf
Department of Electrical Engineering and Computer Science, North South University, Dhaka, Bangladesh

M A Matin
Department of Electrical Engineering and Computer Science, North South University, Dhaka, Bangladesh

Radhika D. Joshi
Department of Electronics and Telecommunication Engineering, College of Engineering, Pune, India

Priti P.Rege
Department of Electronics and Telecommunication Engineering, College of Engineering, Pune, India

Said Benkirane
MATIC laboratory, Department of Mathematics and Computer Science, Chouaïb Doukkali University, Faculty of Sciences El Jadida, Morocco

Abderrahim Beni hssane
MATIC laboratory, Department of Mathematics and Computer Science, Chouaïb Doukkali University, Faculty of Sciences El Jadida, Morocco

Moulay Lahcen Hasnaoui
Computer Science Department, Faculty of Sciences Dhar el Mahraz, Sidi Mohammed Ben Abdellah University, Fez, Morocco

Mostafa Saadi
MATIC laboratory, Department of Mathematics and Computer Science, Chouaïb Doukkali University, Faculty of Sciences El Jadida, Morocco

Mohamed Laghdir
MATIC laboratory, Department of Mathematics and Computer Science, Chouaïb Doukkali University, Faculty of Sciences El Jadida, Morocco

Jhunu Debbarma
Department of Information Technology, Triguna Sen School of Technology, Assam University, Silchar , Assam 788011

Sudipta Roy
Department of Information Technology, Triguna Sen School of Technology, Assam University, Silchar , Assam 788011

Rajat K. Pal
Department of Information Technology, Triguna Sen School of Technology, Assam University, Silchar , Assam 788011

Reza Rasouli
Department of Information Technology, Science and Research branch, Islamic Azad University Kermanshah, Iran

Mahmood Ahmadi
Department of Computer Engineering, Faculty of Engineering, University of Razi, Kermanshah, Iran

Ali ahmadvand
Department of Electronic and Computer, Islamic Azad University, Qazvin Branch, Qazvin, Iran

Hussaini Habibu
Department of Electrical and Electronics Engineering Federal University of Technology Minna, Nigeria

Adamu Murtala Zungeru
Department of Electrical and Electronics Engineering Federal University Oye-Ekiti, Nigeria

Ajagun Abimbola Susan
Department of Electrical and Electronics Engineering Federal University of Technology Minna, Nigeria

Ijemaru Gerald
Department of Electrical and Electronics Engineering Federal University Oye-Ekiti, Nigeria

Indrajit Banerjee
Department of Information Technology Bengal Engineering and Science University, Shibpur, Howrah, India

Prasenjit Chanak
Purabi Das School of Information Technology Bengal Engineering and Science University, Shibpur, Howrah, India

Hafizur Rahaman
Department of Information Technology Bengal Engineering and Science University, Shibpur, Howrah, India

S.Taruna
Department of Computer Science, Banasthali University, Rajasthan, India

Kusum Jain
Department of Computer Science, Banasthali University, Rajasthan, India

G.N. Purohit
Department of Computer Science, Banasthali University, Rajasthan, India

www.ingramcontent.com/pod-product-compliance
Lightning Source LLC
Chambersburg PA
CBHW080645200326

41458CB00013B/4738